OXFORD MEDICAL PUBLICATIONS

The Family in Clinical Psychiatry

The Family in Clinical Psychiatry

SIDNEY BLOCH
*Associate Professor and Reader in Psychiatry,
University of Melbourne*

JULIAN HAFNER
*Associate Professor in Psychiatry,
Flinders University, Adelaide*

EDWIN HARARI
*Consultant Psychiatrist, St Vincent's Hospital,
University of Melbourne*

and

GEORGE I. SZMUKLER
*Consultant Psychiatrist, Maudsley Hospital, London
(Formerly Consultant Psychiatrist, Royal Melbourne Hospital,
University of Melbourne)*

With the assistance of

HENRY BRODATY
*Professor of Psychogeriatrics, Prince Henry Hospital,
University of New South Wales, Sydney*

and

DAVID KISSANE
*Consultant Psychiatrist,
Monash Medical Centre, Monash University, Melbourne*

Oxford   New York   Melbourne
OXFORD UNIVERSITY PRESS
1994

Oxford University Press, Walton Street, Oxford OX2 6DP
Oxford   New York   Toronto
Delhi   Bombay   Calcutta   Madras   Karachi
Kuala Lumpur   Singapore   Hong Kong   Tokyo
Nairobi   Dar es Salaam   Cape Town
Melbourne   Auckland   Madrid
and associated companies in
Berlin   Ibadan

Oxford is a trade mark of Oxford University Press

Published in the United States
by Oxford University Press Inc., New York

A catalogue record for this book is available from the British Library

Library of Congress Cataloging in Publication Data
The Family in clinical psychiatry/Sidney Bloch . . . [et al.]; with the assistance of Henry
Brodaty and David Kissane.
(Oxford medical publications)
Includes bibliographical references and index.
1. Psychotherapy patients–Family relationships   2. Family assessment.   3. Family–Mental
health.   4. Family–Psychological aspects.   I. Bloch, Sidney.   II. Series.
[DNLM: 1. Mental Disorders–etiology.   2. Mental Disorders–therapy.   3. Family
Health.   4. Family Therapy.   5. Models, Biological.   6. Models, Psychological.
WM 31 F198 1994]
RC489.F33F35 1994   616.89'14–dc20   93–45494

ISBN 0 19 262311 7 (Hbk)
ISBN 0 19 262310 9 (Pbk)

Typeset by The Electronic Book Factory Ltd, Fife, Scotland

Printed in Great Britain by
Biddles Ltd,
Guildford & King's Lynn

To our children: Aaron, David, Leah, Simon, Julia, Elana, Rachel, James, and Nuala

Acknowledgements

We are most grateful to our colleagues Margaret Coady, George Halasz, Edmund Chiu, David Ames, John Buchanan, Julie Jones, Julian Leff, Ivan Eisler, Paul Brown, and Paul Holman who read sections of the manuscript and offered valuable suggestions. We are especially indebted to Henry Brodaty who wrote the chapter on dementia and to David Kissane who was the principal author of the chapter on grief. It is a pleasure to record our gratitude to Robin Skynner for his generous foreword, and of our appreciation of his pivotal contribution to family psychiatry.

We would like to thank Helen Smart, Ann Regan, Julie Larke, Pauline Ng, and Kathy Madden for their help in preparing the manuscript. Our gratitude also to Gillian Hiscock and her colleagues at the Victorian Mental Health Library and Sandra Russell and her colleagues at St Vincent's Hospital Medical Library for their magnificent efforts in locating bibliographic material. We thank Gertrude Rubinstein for her painstaking proof-reading. Sidney Bloch was a visiting research fellow at the Australian Institute of Family Studies in the first half of 1992; he was made to feel most welcome by its director, Dr Don Edgar, and his staff and given great support for the project. Finally, it is our pleasure to thank the staff of Oxford University Press for their splendid support and encouragement.

Foreword

Robin Skynner

From time to time, new knowledge emerges which is not only practically useful in itself, but which also throws a new light on all that has been discovered before and renders it more meaningful. For me, one example was the papers which appeared in American journals during my training in the 1950s, describing experiments in bringing whole families together to discuss the problems of the referred patient. I did not feel bold enough to try this approach myself until 1962, but the ideas presented acted like a yeast on everything that I had learned before and when I finally began similar experiments myself during my first consultant appointment in child psychiatry my life was profoundly changed through the unexpected widening of scope of psychotherapeutic intervention and the elimination of waiting-lists it made possible.

I was privileged to play some part in the spread and development of training in these new methods, which were soon widely employed throughout the child psychiatric services in the UK. But it has remained something of a mystery why so little equivalent interest has been aroused among psychiatrists working with adult patients. It is not because the methods of family interviewing are less effective when all the participants are adult, for in many ways it is easier when everyone is accustomed to communicating verbally, and the colleagues and I who taught and demonstrated these techniques in adult departments of the Maudsley Hospital found them to be widely applicable and often particularly useful with some severe disorders resistant to previous treatments.

One factor contributing to the slow take-up is no doubt the fact that many texts by family therapists are not user-friendly to their general psychiatrist colleagues, and have developed a mystifying and alienating jargon to rival that of psychoanalysis. There has long been a pressing need for a text to remedy this, and I was delighted to discover how well this present volume meets what is required. The range and depth of the co-authors' own psychiatric experience ensures that it focuses mainly on the kinds of case which occupy the bulk of the general psychiatrist's time, showing how our increased knowledge of family interaction can suggest ways to complement and support established forms of therapy, enlisting the co-operation and help of the family, as well as securing the compliance of the referred patient, in achieving treatment goals.

The book's other merits will also inspire confidence in this potential readership. Research findings on each topic are well-documented and referenced, particularly those assessing the efficacy of family interventions.

It does not attempt to be a new manual of family therapy techniques, but different approaches are clearly outlined with ample guidance for further reading, making it an ideal introductory text for its purpose. Strategic and systemic methods are included, as well as psycho-educational interventions and those influenced by psychoanalytic or behavioural ideas, while the authors are attentive to ways in which the traditional separate interviews with relatives of nominated patients can become more meaningful. Above all, conjoint interviewing of whole families is throughout regarded as one additional tool among many, most useful when treated as such and combined as necessary with more established treatments, thereby avoiding presenting them as methods claimed to surpass or aimed to replace those with which colleagues are more familiar. I am sure that the latter attitude on the part of many family therapists has done much to arouse resistance unnecessarily and delayed the wider acceptance of this new knowledge.

All family therapists will gain a broader and deeper understanding of their work by reading this book. And I am sure that no open-minded general psychiatrist (and also other mental health professionals) reading it can fail to find his or her work empowered and made more interesting and enjoyable by the information it so succinctly and readably conveys.

Preface

In 1965, the World Health Organization published a volume entitled *'Aspects of family mental health in Europe'*. The preface began thus:

The study of mental disorders began as the study of individuals, and only gradually did it become the study of individuals in the social environment. As a result of this approach, it has become increasingly apparent that the mental health – or ill health – of each member of a family is inextricably bound up with that of other members, influencing it and being influenced by it both favourably and unfavourably.

The ideas expressed so succinctly in this paragraph are at the core of this book as will become evident from a perusal of the chapters that follow. But we need to elaborate on how we understand the role of the family in clinical psychiatry and we do that in this preface.

THE 'GENUINE' MEDICAL MODEL

Psychiatry operates within the framework of medicine in general and thus tends to apply the medical model. If this model is deployed appropriately, distinct advantages accrue. Rawnsley (1988), for instance, avers that the clinician is better placed in his encounter with a patient through an appreciation of the relevance of a range of key factors:

The medical man or woman by virtue of the training they have received are able to draw on a biological standpoint, a psychological standpoint, and a social standpoint all of which should come together in the assessment of the clinical problem. The imputation of the term 'medical model' is that it is to do only with organic factors, drugs and a mechanistic way of looking at patients. This is quite wrong. The medical model, as I define it, brings in all angles from the biological to the social. A good doctor uses a complex, elaborate point of view, in assessing a clinical problem. That is the medical model in my opinion. A unique perspective . . .

George Engel (1977) has provided the most influential argument for such a triadic view since his promulgation more than a quarter of a century ago of a new approach to the traditional medical model – relabelled by him as the biopsychosocial approach in order to stress its three core dimensions. By drawing upon Systems Theory, Engel (1980) has suggested persuasively that this broader view enables the clinician: 'to extend application of the scientific method to aspects of . . . patient care heretofore not deemed accessible to a scientific approach'.

In this book we stand firmly behind Rawnsley and Engel in supporting their notion that equal weight should be given to the biological, the psychological, and the social in clinical practice and, as importantly, to their interrelationships. Features stemming from each of these domains need to be addressed in every patient, whatever the nature of the presenting clinical problem. The biological and psychological dimensions are more easily understandable. Consider the biological dimensions; we may detect links between particular physiological states and psychiatric symptoms (for example, thyrotoxicosis) or between structural change and symptoms (for example, cerebral tumour) or between genetics and psychiatric conditions (for example, Huntington's Disease). Similarly, a group of psychological models such as Psychoanalysis, Cognitive Theory, and Personal Construct Theory provides a reasonable foundation for conducting psychiatric assessment and thus achieving an understanding of psychological processes relevant to the patient's problems.

The social dimension of the medical model by contrast has proven conceptually elusive. Over seventy years ago the legendary Adolf Meyer (1921) highlighted the challenge to clinicians to grapple with the social aspects of their patients. That challenge remains pivotal for the contemporary practitioner. The social dimension involves diverse forces such as the effects of unemployment, racism, sexism, migration, socio-economic status, ethnicity, adverse life events, support networks, and the like. In our view, however, it is wise to start at the most salient and fundamental level of the patient's social world, the irreducible unit of society, namely the family – including the nuclear family, the extended family, and the family of origin – since it is most commonly in this context that a person presents clinically. Invariably, interactive effects occur in that the patient impinges on his family and family members upon him.

IMPLICATIONS FOR CLINICAL PRACTICE

If we regard the social dimension as seriously as the biological dimension and the psychological dimension and specifically use systems thinking, certain implications follow for the clinician. Among the chief ones are:

1. The clinician should take a family history in a way that highlights the functioning of the family system (whether nuclear, family of origin, or extended). As we point out in Chapter 3, this is best achieved by the drawing of a comprehensive family tree in that it provides not only a graphic display of the system's members but also of their interactions. Far from being a merely factual representation, the tree enables the family dynamics to spring to life.

2. The clinician may need to conduct a family interview to gather information which cannot be elicited from the patient alone and, as

importantly, to observe directly the function of the family in the here-and-now.

3. Whether such an interview occurs or not, the clinician should as part of the formulation add a systemic dimension, derived from the assessment procedure, which is concerned with the functioning of the family (and possibly other social groups) and its association with the patient's presenting clinical features.

4. Additional family interviews as part of a programme of family therapy may or may not ensue. If they do, a cardinal purpose is to explore the systemic dimension which in turn may pave the way for potential alterations in the system and corresponding change in the patient and other family members.

CAVEATS

Adding this form of social assessment to the diagnostic and therapeutic process is not without its difficulties. We certainly would not wish to obscure the real hurdles standing in the way. These have been described vividly by Jay Haley (1975), and the reader is referred to his paper. In summary, the following constitute the key hurdles:

1. A paradigmatic shift is required in that conceptual thinking of a new type is called for; this may make a major demand on a clinician who has been trained to view the problem as located in the individual patient.

2. If the clinician works as part of a multidisciplinary team, a shift is necessary in the orientation of all its members. Problems may arise when some regard the new paradigm as more acceptable and appealing than others.

3. There is little doubt that when a family is brought into the clinical arena, the interventions that follow are necessarily potent. Families commonly report that they have never met and talked in this particular way. Secrets may be revealed; candid feedback may be offered between members. This form of work clearly calls for consummate skill and experience, and certainly cannot be executed without adequate training.

THE PURPOSE OF THE BOOK

'*The family in clinical psychiatry*' is an attempt to aid the clinician to overcome these snags and to realize the full potential of the biopsychosocial model. We have approached our task in what we suspect is an unprecedented way. Rather than tread onto the terrain of the family therapist and in so doing jeopardize the cogency of all three dimensions of the biopsychosocial model, we adhere to a generally conventional clinical approach with which all mental health professionals are familiar. Indeed, we assume a basic knowledge of clinical psychiatry in the reader.

We need to stress that we have not prepared a text on family therapy and that our volume barely resembles the myriad of texts written on that subject. In a sense, we return to fundamental issues which precede the themes inherent in family therapy *per se*. Unfortunately and paradoxically the rapid burgeoning of family therapy practice since the 1960s has resulted in a state of professional confusion with polarization of viewpoints; one is either a proponent of systems thinking, and hence of the widespread application of family therapy, or one is a conservative sceptic of that position.

The question of how the family should be viewed in the context of the conventional clinical syndromes is far from sorted out. For example, some clinicians erroneously assume that assessment of the family's role *vis-à-vis* a psychiatric condition is axiomatically an indication for 'formal' family therapy. On the other hand, clinicians whose mode of practice has been exclusively with the individual patient may fail to appreciate the relevance of family forces or even be wary of them. A study by Cotterell (1989) of British psychiatric trainees demonstrated how few of them attended to family issues in the assessment and subsequent management of their patients.

A striking consequence is that despite the fast growing movement of family therapy, a correspondingly widening gap between the language, concepts, and practices of the general mental health professional and those of the family therapist has developed.

We have sought to bridge this gap and to address the doubts of the sceptics on both sides by demonstrating that an integrated clinical stance is not only feasible but indispensable if the biopsychosocial model is to be properly implemented. We hope the book will serve as a scholarly, non-doctrinaire guide for clinicians and facilitate their adoption of a comprehensive diagnostic and therapeutic approach, more particularly by describing how family factors relate to the major psychiatric disorders in terms of assessment and general treatment.

THEMES INCLUDED AND EXCLUDED

The first two chapters cover theoretical and research aspects of the family that are pertinent to psychiatric practice. The concluding chapter deals with the important topic of ethics as it relates to the patient and his family. The middle section focuses directly on clinical syndromes. In this regard, we have opted to limit ourselves to the main areas of clinical psychiatry where the role of the family is substantially relevant. We are aware that clinical life is not so simple and in particular that co-morbidity is common and that psychiatric diagnoses are not always static entities. But, for heuristic purposes, we have proceeded to deal with syndromes as if they were independent of one another and occurred singly.

In the case of the clinical chapters, our goal has been to distil contemporary knowledge about the relationship between family factors and particular clinical conditions. Rather than providing an exhaustive review of the literature, we have focused our attention on the most noteworthy matters in the case of each condition. Thus, for example, we concentrate in schizophrenia on family factors in relation to relapse whereas in dementia the central theme is the role of the family caregiver. We have taken care in this task to dispel existing myths and to prevent their substitution by new ones. In this spirit, therapeutic *implications* rather than recommendations are included in these chapters. Our premise is that current knowledge permits us only to raise potential ideas concerning treatment. The subject is simply too new to extend beyond this first step. In offering implications we should reiterate their chief source in extant knowledge rather than in our own whims and idiosyncratic judgements. On the other hand, we have been obliged to interpret available research and clinical data, a process necessarily influenced by our own collective clinical experience. So, the reader who seeks a 'how to' account will not discover it here. Contrariwise the reader will, we envisage, be better equipped to handle texts on family therapy by using them with a more critically analytic eye.

Our book does not aim to be encyclopaedic (we have already alluded to selective rather than exhaustive bibliographies). It will be especially noted that child psychiatric syndromes are not addressed specifically. The reasons for this are simple: (1) five of the six contributors work in the adult clinical setting and are thus relatively unqualified to comment on child psychiatry and the family; and (2) the book would be unwieldy were we to enter the area of child psychiatry (incidentally, there is ample scope for separate volumes on the role of family factors in child psychiatry as there are for other specialities including psychogeriatrics, mental handicap, and psychotherapy).

CONCLUSION

The full potential of the medical model with its biological, psychological, and social components emphasized equally remains to be realized. We believe that psychiatry has been presented with an excellent opportunity to accept this challenge. We hope that our book will contribute in some way to that process.

Finally, we should mention that our use of the male pronoun in the text, except where obviously inappropriate, is merely to avoid cumbersome repetition of both male and female pronouns.

REFERENCES

Cotterell, D. (1989). Family therapy influences on general adult psychiatry. *British Journal of Psychiatry*, **154**, 473–7.

Engel, G. (1977). The need for a new medical model: a challenge to biomedicine. *Science*, **196**, 129–36.

Engel, G. (1980). The clinical application of the biopsychosocial model. *American Journal of Psychiatry*, **137**, 535–44.

Haley, J. (1975). Why a mental health clinic should avoid family therapy. *Journal of Marital and Family Therapy*, **1**, 3–13.

Meyer, A. (1921). The contribution of psychiatry to the understanding of life problems. In *The commonsense psychiatry of Adolf Meyer* (ed. A. Lief) 1948, pp. 1–15. McGraw-Hill, New York.

Rawnsley, K. (1988). Interview with Kenneth Rawnsley. *Bulletin of the Royal College of Psychiatrists*, **12**, 49.

Melbourne and Adelaide   S.B.
March 1994   J.H.
                                                                E.H.
                                                                G.I.S.

Contents

Part I Theoretical and research aspects

Part II Clinical aspects of the family in psychiatric practice

Part I

Theoretical and Research Aspects

1

The well-functioning family

The family in clinical psychiatry focuses principally on the relevance of family factors in the aetiology, treatment, and prognosis of the major psychiatric conditions encountered in clinical practice. Given that in these circumstances family function is in one way or another often disturbed and that a key objective therefore may be to intervene therapeutically at the level of the family, we need to be aware of the 'normal', 'healthy', 'functional', or 'well-functioning' family. Our assumption is that promoting the family's function is a desideratum in the overall therapeutic approach to the presenting patient.

Appropriate terminology is elusive when discussing the 'healthy' family, hence the quotation marks above. Moreover, identification of universal criteria to establish whether a family is 'healthy' is problematic. Even more so is the attempt to signify what constitutes a 'normal' family. Indeed, Don Jackson (1977), a pioneer of the family approach in psychiatry, concluded that there is no such thing as a normal family. Instead, the family, he averred, is a relational network where the beliefs and behaviours of its members fit or do not fit comfortably. Tolstoy's often quoted opening sentence to his great novel, Anna Karenina, that, 'All happy families are alike but an unhappy family is unhappy after its own fashion' may be an oversimplification.

Jackson may have been wise to desist from defining the normal family but his remedy remains problematic. By what authority can we determine whether the family's beliefs and behaviours fit or do not fit comfortably? In the final analysis, there is no such authority, only the perspectives of each family member. A father's insistence that, according to his family's tradition, he should occupy the role of sole family leader may be unopposed by his wife and children but the ostensible 'fit' may be little more than an expedient measure by them to maintain a measure of stability. Behind the compliance may lurk a seething resentment that the status quo is grossly unjust.

The core issue emerges obviously enough: the pathway to a definition of what constitutes a normal family is full of snags and snares. Perhaps the most trenchant critical analysis of these matters has been generated by the feminist movement. Deborah Luepnitz (1988) is a most eloquent spokesperson for a position which avers that the apparently normal family

is one typified by gender inequality with resultant unfairness experienced by
the wife (and also by her daughters). She reminds us of the central place of
value judgements when stating:

The question of what one decides to term adequate or functional and what one
decides is instead a serious social problem is a question of values or ideology. And
indeed the most transformative insight that a feminist perspective can offer . . . is
the realisation that ideas do not fall from the sky; they are artifacts constructed
by people whose thinking is never ideologically impartial.

Luepnitz in her well documented volume on feminist theory in clinical
practice proceeds to demonstrate how many renowned authorities in the
field of family therapy have been prisoners of their own ideology with a
particular citation of conservative male figures who believe that women
should serve their husbands and family and 'produce well-behaved and
socially compliant children'. She is particularly critical, for instance, of
Nathan Ackerman (see later in this chapter), one of the 'fathers of family
therapy' for a view which embodies gender differences of a flagrantly
patriarchal type. Thus, women should not denigrate their traditional
mothering role or envy the male, lest they become burdened with anxiety
about their 'incompleteness as women'; men should wear the trousers in
the family whereas women should be passive.

Luepnitz does concede that values inevitably relate to a temporal and
cultural context. Ackerman shaped his ideas during the 1950s in the USA,
a conservative era in American society and prior to the feminist revolution
of the 1960s and beyond. But the point remains, and her commentary
demonstrates this well, to define the normal family is a subjective matter
and relative to a socially constructed reality.

An option to obviate this difficulty of definition is to concentrate
on aspects of family function which appear disturbed with the limited
goal of modifying these alone. We could designate this the clinically
pragmatic model.

Jay Haley is a notable representative of this view (1963) with his
'strategic' therapeutic approach to the family. An alternative option, not
uncommon among family therapists, is to 'perturb' the family in such a way
that its members' curiosity is aroused in terms of change but the precise
nature of that change is regarded as less salient than the fact of change itself
(Jenkins 1989). A variation on this option is to construe the family's position
as one of 'stuckness' and to work towards liberating the group in order that
its members might be better equipped to re-establish a pattern of growth
and development. Again, it is a matter for the family to determine its own
direction.

We may readily note that all three options do not require a precise
definition of the well-functioning family, this making the clinician's task

considerably easier in terms of setting therapeutic goals but not necessarily overcoming the above-mentioned issues surrounding values.

And yet common sense would suggest that change is predicated on the notion that the clinician and/or the family have clues as to what qualities enhance its welfare. Fortunately, a body of knowledge has evolved since the 1960s whereby we can identify intrinsically advantageous and beneficial aspects of family life. Although, it is remarkable how scanty has been the specific research on the healthy family, with interest much more focused on families in difficulty, there are a few outstanding investigators who have devoted considerable effort to determine what constitutes normal function.

Knowledge in this context stems from three chief sources: (a) theoretical formulations pointing to desirable family qualities; (b) the findings of empirical research, both clinical and experimental (see, for example, the work of Reiss 1981; Olson *et al.* 1979; Epstein *et al.* 1982, and Wynne *et al.* (1982); and (c) through extrapolation from therapeutic encounters. The rest of this chapter is devoted to a distillation of this knowledge. Not unusually, theorists, researchers, and therapists pay little heed to one another's contributions (Eisler *et al.* 1988). Our task here is to consider the literature in all three areas although we devote a separate chapter to research on models of family function (see Chapter 2). We also attend to one other aspect which has a marked influence on family function, namely the family life cycle.

Let us begin with a note regarding family structure. Delineating a typical form of the family has become impossible in recent years. The extended family was generally prevalent until the twentieth century when, certainly in Western societies, it came to be replaced by the nuclear family, i.e. the parental couple and their offspring (Shorter 1975). Latterly, we have witnessed the proliferation of a range of other family forms (Luepnitz 1988). They include the divorced or separated family in which any children are customarily but not invariably cared for by the divorcee mother (we need to recall that between one in four and one in three marriages breaks down); the single-parent family in which a mother rears her offspring without having been part of a marital or other intimate bond; a reconstituted or blended family in which previously married parents form a new relationship and are joined by any children of either parent. Challenging the very concept of what defines a family is the newly emerging pattern of long-term homosexual partnerships as well as the mothering of children by homosexual women, either in a relationship or as a single parent (Janosik and Green 1992).

Other forms of family are pertinent in terms of their function being strongly influenced by a dominant characteristic such as childlessness (we should remember that 10–15 per cent of couples are infertile (National

Bioethics Consultative Committee 1991), the presence of adopted children, and the absence through death of one parent. Finally, exceptional family forms are encountered in certain social groups. The kibbutz or communal settlement in Israel is a good example; here, children are reared collectively allowing parents more flexibility in contributing to the welfare of the social group overall (it is noteworthy however that this arrangement is undergoing radical transformation in many kibbutzim, the traditional nuclear family pattern emerging as the common alternative).

FAMILY FUNCTION

Whatever the family form, we may still consider its function more generally. But first, pause for a moment to reflect on the history of the family as a social institution (Shorter 1975). The family has in fact endured for centuries, a period during which many other social organizations have either disappeared, fragmented, or undergone radical change. What underlies this enduring quality? When Santayana avowed that the family is one of 'nature's masterpieces', he was presumably alluding to its inherently sociobiological function – the continuation of the species through procreation as well as the maintenance of its welfare through the drive to nurture its offspring in the face of omnipresent risks confronting the highly vulnerable and dependent infant and young child.

Nathan Ackerman (1958), a pioneer of family psychology, has put it well when asserting that:

None of us lives his life alone. Those who try are foredoomed; they disintegrate as human beings. Some aspects of life experience are, to be sure, more individual than social, others more social than individual; but life is nonetheless a shared and sharing experience. In the early years this sharing occurs almost exclusively with members of our family. *The family is the basic unit of growth and experience, fulfilment or failure* [our italics]. It is also the basic unit of illness and health.

Ackerman has also helpfully identified the social purposes of the family:

(1) to provide basic necessities in order to sustain life as well as protect from danger;

(2) to provide a sense of cohesiveness;

(3) to offer family members the opportunity to develop a sense of identity;

(4) to provide a suitable context for sexual maturation;

(5) to promote socialization into various social roles (to this should be added the promotion of a moral sense in the offspring – see Piaget (1977) and Kohlberg (1981)); and

(6) to encourage each member's creativity.

We shall consider some of these purposes in more detail later but now turn to features typifying the well-functioning family. We select this terminology rather than alternatives like normal, happy, and functional, since it indicates more explicitly what we are trying to convey and, to a degree, deals with the risk of the researcher imposing his personal values on his inquiry. The categories listed below are inevitably arbitrary in part, reflecting our own understanding and interpretation of the relevant literature coupled with our experience of treating disturbed families. There are indeed several ways to cut the pie; our method is but one approach. A major hurdle in this pursuit is the variegated nature of the contributions of theorists, therapists, and investigators together with the absence of an overriding model which embraces all dimensions of family function. Our categories are as follows:

(1) clear delineation of the roles occupied by family members, and their associated tasks and responsibilities;

(2) relationships between members which are warm and affectionate but at the same time respectful of the separateness of each participant;

(3) the joint appreciation and acceptance of explicit rules to regulate the family's behaviour and conduct;

(4) the capacity to face challenges and withstand changed circumstances, either predictable (inherent in the family life cycle) or accidental, and the avoidance of domination by past 'scripts' or 'myths';

(5) clear, open, and direct communication between members, including tolerance of conflict and a readiness to grapple with differences when they occur;

(6) an ability of the nuclear family to relate readily with other significant social groups such as the extended family, children's schools, friends, work colleagues, and neighbours.

We note the potential overlap between many categories but we have teased them out in this way in the interests of clarity and conceptualization. We now deal with each in turn.

FAMILY ROLES

In the well-functioning family, members occupy particular roles (or sets of roles) of which they and the rest of the family are aware. The qualities of the roles relate to tasks, responsibilities, expectations of others, and individual rights. Thus, parents are expected to assume roles of leadership thereby taking on responsibility (duties may be a relevant term here) for the nurturance and care-giving of their children. This is highlighted by Minuchin (1974) who asserts in his structural model of the family that parents should be accorded executive functions, occupying the top positions in the family hierarchy. Alongside this, parents are accorded power to wield in order that they might fulfil their obligations. Such differentiation of roles between parents and children is obvious but other distinguishing patterns may also apply. For example, an older sibling may be assigned the task of serving as a temporary care-giver of a younger child when both parents are unavailable. This also illustrates a highly relevant point, that of flexibility of role with the periodic assuming of non-customary roles when necessary.

Other roles may be less instrumental, for example, a buoyant, lively youngest child may imbue his parents with a playful spirit and so elicit the child element in them, promoting a sense of fun in the household. Another child may, through his curiosity and eagerness to learn, set off a similar spark in one or both parents.

In addition to clarity of roles, other features are required for optimal family functioning. Family members should feel secure and at ease in their roles gaining respect from one another for their contributions but yet remain prepared to experiment with other, less familiar ones. The roles should contribute to self-esteem in the sense of members experiencing their contributions as meaningful and relevant. Contrariwise, no member should feel exploited by being allocated a role which is inappropriate or unfair.

WARM AFFECTIONATE RELATIONSHIPS ASSOCIATED WITH MUTUAL RESPECT FOR INDIVIDUALITY

The issue of relationships between members is multidimensional, and several theories are available to account for this. The core feature however revolves around the provision within the family of a balance between distance and closeness (see Minuchin 1974). For example, marital partners relate intimately to each other but not at the expense of the need and wish to retain their individuality and autonomy. Helm Stierlin and his colleagues (Simon *et al.* 1985) has described this aptly with their term 'co-individuation' (co-evolution is another useful term in this context). The implication is that each member is separate from the rest of the family although retaining affectionate and mutually respectful ties.

Serving as a contrast to co-individuation is Minuchin's concepts of disengagement and enmeshment. In the former, family members are so distinct from one another that cohesiveness and a sense of belonging are lacking or entirely absent. An enmeshed family by comparison fails to provide for individual autonomy and independent growth; the family can be described as a single, undifferentiated 'ego mass' with little or nothing to distinguish between members (Bowen 1961).

The concept of boundaries is also pertinent. Applying the metaphor of a membrane to represent the boundary, in tandem with General System Theory (GST) (Bertalanffy 1968), we can depict the family as a system comprising a number of parts but as a system which because of its multiple relationships is greater than the sum of those parts. The membrane in a balanced, connected family is sufficiently permeable to permit intimate and reciprocal age-appropriate relationships to occur. Excessive permeability typifies the enmeshed system, diminished or absent permeability the disengaged system. Using GST further, we note that the concept of boundaries also applies to sub-systems such as parents on the one hand and children on the other. Again, relationships between sub-systems are influenced by the type of boundary separating them. In a well-functioning family, for example, parents agree about how much 'supervision' they need to provide for their children, the level and the form depending to a large degree on the stage of the family life cycle (see below).

A further concept regarding family relationships is relevant, derived from the work of Gregory Bateson (1972). At least three contrasting forms occur, each having a recognized function according to the requirements of the relationship in question. An additional desideratum is that these forms of relationship are not fixed but alter to suit changing circumstances.

In the *symmetrical relationship* each partner (or group) exchanges a similar type of behaviour with the other, for example, both parents express emotional support to each other; a destructive symmetry by contrast would be the equal hurling of disrespectful innuendos between parents.

A second form of relating is characterized by *complementarity*, that is, the protagonists relate unequally with the behaviour of one eliciting an opposite response from the other. Complementarity may be applied adaptively or destructively. An instance of the former is the circumstance in which, say, a husband in the midst of a stressful dilemma transiently relies on his wife for support and guidance, the wife adopting the caring role voluntarily and empathically (this reminds us of a previous point, the flexible occupancy of roles). Furthermore, an implicit or explicit shared understanding prevails whereby this new form of relating is regarded as apt in the circumstances but not enduring beyond the resolution of the difficulty encountered by the husband. Therefore, flexibility in allowing various forms of relating is accepted with the family acting in appropriately adaptive ways to accommodate changes in relational patterns (recall Jackson's comment on the family's comfort with its relational network).

The third relationship pattern is the *reciprocal*, a blend of the first two, wherein each partner manifests similar patterns but rather than symmetrically they do so asymmetrically, for example, one partner assumes a dominant posture and the other a dependent one but these roles are subsequently reversed; a sequence may follow whereby each partner adopts the same pattern but not concurrently. Thus, at no stage will a partner act dominantly or dependently while the other does so. This makes for adaptive relating given the flexibility involved and the propensity for members to take up contradictory positions according to the family's particular needs. Failure of reciprocity occurs if complementarity is rigidly adhered to, thus precluding a possible exchange of positions.

A final dimensions of relating within the well-functioning family concerns three linked dynamics: alliance, coalition, and scape-goating. The first is perfectly adaptive, the latter two an indication of disturbance. In an *alliance*, two or more members unite to promote a shared interest, need, or purpose. Father and children for instance may pair up in order to prepare a surprise party for mother's birthday. The grouping has no sinister implications and is at no one's expense. The opposite is the case where, say, a father recruits his children to his 'camp', thus forming a *coalition*, in the face of marital discord and imminent family fragmentation. The purpose here is to exploit, unjustifiably, the rest of the family in what comprises a sub-systemic problem, one belonging to the marital pair. *Scape-goating* based on the defence mechanism of projection, is equally destructive of relationships inasmuch as one or more members are unjustifiably portrayed in critical or hostile terms accompanied by corresponding reactions of rejection, even ostracism.

THE APPRECIATION OF THE PLACE FOR RULES

Don Jackson (1965*a*,*b*) was among the first to point out the centrality of family rules when he emphasized their utility as a stabilizing force. Family rules, for Jackson, parallel the interactional pattern prevailing among members; in part, the rules define the relationships. For instance, a rule requiring that a husband deal with decisions pertinent to the external world whereas his wife manage domestic matters could reflect a complementary relationship in which the husband is viewed by the family as substantially more powerful than his spouse.

Rules in the well-functioning family are typified by a number of note-worthy qualities. Firstly, rules settled upon by the family suit the needs and interests of all members; no member is disadvantaged or discriminated against. In this regard, Boszormenyi-Nagy's (Boszormenyi-Nagy *et al.* 1991) concept of 'relational ethics' is relevant (see Chapter 14). Such ethics serve to accomplish an 'equitable balance of fairness' among members so that their fundamental interests are seriously considered by the whole family. The reciprocal process is based on trust, specifically the family's realization that trustworthiness is itself an ethical accomplishment. Rules ultimately agreed upon by the family are in this way governed by a principle of justice.

Secondly, rules aid the family in affecting the conduct of members in diverse ways – from regulating the distribution of power among the sub-systems to providing a basis upon which conflict can be tackled and resolved.

Thirdly, rules are either prescriptive (that is, explicit) or implicit but usually applied automatically (that is, 'this is how we do it in our family'). In both cases however they are sufficiently clear to be appreciated by the family as appropriate and necessary for its functioning.

Fourthly, the rules are actually applied when circumstances require it. There is therefore consistency and predictability regarding implementation. But, leeway also exists for flexibility when particular situations warrant it. In such circumstances, however, the rule-breaking needs to be justified.

Fifthly, rules are not set in stone. They come under scrutiny for possible revision or even disposal. Change is especially pertinent as the family moves into another stage of its life cycle, for example, rules suitable for a primary-school-aged child will necessitate major amendment when he embarks on adolescence.

Finally, the rules contribute substantially to the management of conflict, paving the way for negotiation between 'combatants'. The clearer the rule, the more effective its utility in the process of negotiation. A confirmed infringement may call for sanctions but these will have been previously identified and agreed upon by the parents, that is, the meta-rule is who decides. A simple example is as follows: if an adolescent has returned home in the early hours of the morning instead of at an agreed upon time, the sanction, defined as part of the rule, of forbidding other social outings for a prescribed period may be reasonably applied.

A final point relates to the children sub-system requiring an appreciation of salient rules. A study by Dunn and Mann (1985) on the social development of a child during the second year of life reveals interestingly how this comes about. In the context of conflict, the child becomes progressively more aware of the family's rules: the family concurrently makes adjustments to the child's emerging contribution to rule-setting. The research also demonstrates the younger child of a two-child sibship requiring more understanding of the rules whereas both mother and the older sibling are able to refer more directly to the relevance of particular rules in circumstances of conflict.

THE CAPACITY TO FACE CHALLENGES AND TO DEAL EFFECTIVELY WITH PROBLEMS

As we shall see in the section on the family life cycle, the family cannot avoid the negotiation of developmental change points which are an inherent aspect of that cycle, for example, the birth of the first child, the birth of subsequent children, deaths, marriage of an adult child, and the like. The family is similarly prone to less predictable but commonplace accidental (or 'critical') stresses such as illness, a motor car accident, loss of job, failure in examination, and one hundred and one other life events.

In the face of the ineluctable, the family needs to deal with these assaults on its wellbeing. The requisite abilities are many, including clear communication, role differentiation, cohesiveness, mutual support and the sharing of burden, and acceptance of family rules. Since these qualities are covered elsewhere in the chapter we concentrate here on the 'essentials' of effective coping with problems in all their forms.

We can confidently commence with the premise that in the well-functioning family the members themselves are the primary resources for managing problems. An additional 'in-built' factor is the notion that all systems, including the family, tend towards stability (synonyms here are 'equilibrium' or 'homeostasis') that is, in the face of change which generates disequilibrium, the family reacts directly to re-achieve stability, albeit on a new plane. Moreover, the family is open to stimuli through its semi-permeable boundary with the external world and is therefore accustomed to the fact of a constantly altering environment.

The more behaviourally oriented theorist posits the following sequence in a family's effective problem-solving (see, for example, Falloon *et al.* 1984 and Epstein *et al.* 1982). Identifying the problem and its nature and informing relevant members (young children may for instance be spared) constitutes the initial step. Alternative options are then formulated through a 'brain-storming' process involving, *inter alia*, a calculation of their relative benefits and costs. One option is then selected and corresponding action taken. A monitoring procedure is implemented which culminates in a review of the action's overall efficacy. Any subsequent decision-making and action are contingent on the findings of the review.

In more general terms, well-functioning families can discern, through their own experience, problem-solving strategies that have worked effectively or that have failed; they act accordingly by ditching the latter and applying the former or alternatively by creatively exploring the possibility of other, previously untested approaches (Weakland 1974).

This sequence is derived from the therapeutic programmes of such practitioners as Epstein *et al.* (1982) and Falloon *et al.* (1984) but it is not unlikely that the well-functioning family deploys similar schemes even if not as systematically or explicitly as these therapists would assert.

Yet another quality commonly cited as pertinent in the family's tackling of problems (and much more beside) is adaptability. We shall have more to say about this topic in the following chapter (see Olson *et al.* 1979). Suffice to comment here, adaptability refers to the family's capacity to withstand changes required by the emergence of new circumstances.

The adaptive family values stability as much as any other but in the face of a challenge or problem risks the necessary changes, giving up the familiar and the secure in the interest of reachieving stability even if of another form, so-called 'morphogenesis'. Moreover, as part of its capacity to react adaptively the family assumes flexibility in relation to family rules, roles and structure, the assimilation of new ideas from whatever source, and the accommodation of responses by various members to the problem confronting them.

Apart from the here-and-now approach to problem-solving as mentioned above, a family also draws on past experience of grappling with similar problems. In this respect, the important topic of the family 'script' is pertinent (other comparable terms include family 'myth', 'world view', and 'paradigm').

Reiss (1981) has demonstrated experimentally how every family evolves a coherent view of the world, the 'family paradigm'. This is confirmed by members' awareness of their paradigm in terms of assumptions and beliefs they share which in turn enables them to define themselves as a family group. The paradigm may also manifest itself through rituals, routine behaviour, or patterns of problem-solving (the latter particularly relevant to the issue under discussion).

John Byng-Hall (1988) utilizes the metaphor of the family script. Just as actors follow the script of a play or film, so does a family behave according to its shared beliefs. Both marital partners bring their own scripts into their new relationship and, with sufficient overlap between them, the couple accomplishes a new equilibrium. The emergent family script prescribes the pattern of family interaction in specific contexts. The well-functioning family, according to Byng-Hall, is sufficiently resilient to permit of a 'corrective' script, that is, the family attempts to correct for past mistakes rather than persist with a 'replicative' script which precludes the possibility of change. Thus, for instance, a family that has customarily depicted itself as always failing in the face of challenge or adversity is constrained in its response by an expectation of repeated failure, whereas a family with a coherent perception of itself which incorporates a sense of solidarity and collective confidence in its ability will tackle a problem with a realistic measure of optimism.

CLEAR, OPEN, AND DIRECT COMMUNICATION BETWEEN MEMBERS

Communication within a family has been emphasized by several commentators in the family psychology and family therapy fields, the consensus emerging that sharing of information, feelings, and attitudes is of paramount importance (Satir 1967). In the well-functioning family communication has well-defined qualities. Members express themselves openly, directly, clearly, and congruently. They address one another without need for intermediaries and avoid ambiguity. Moreover, the content of their message is accompanied by corresponding emotion and other features of non-verbal communication. The critical issue of congruence stems from the work of Watzlawick and his colleagues (1967). They differentiate between two forms of communication, digital and analogic. The former represents the actual content of the message conveyed; the latter, reflecting the form of relationship between the people communicating, refers to aspects like tone, inflection, and cadence as well as non-verbal signals like gesture, stature, facial expression, and other bodily movement. The analogic level is pivotal inasmuch as it defines the relationship between those communicating.

In the well-functioning family, members not only convey content explicitly but also manifest corresponding feeling and other analogic qualities, obviating any potential for misunderstanding.

Communication also necessarily deals with conflict. Rather than evade matters awkward or difficult, the well-functioning family 'fights well', but also 'cleanly' according to a set of explicit rules. Parents for instance are sanctioned by dint of their authority to discipline a child if he contravenes a rule applicable to the children. Similarly, parents may exchange heated views about a family matter but with the proviso that the dispute is not at the personal cost of either protagonist. Thus, both are entitled to express their views, even forcefully. Inherent in this process are mutual respect no matter how divergent the viewpoints and the potential for negotiation and compromise.

All social groups are necessarily buffetted by differences between their members. The family is no exception. Indeed, given the extended period in which members live so intimately, conflict is inevitable. The overly placid family is invariably constrained in not possessing the licence to express itself fully and spontaneously lest harm ensues or criticism levelled for disturbing the 'pax familia'. A family without experience of conflict is bound to be adversely affected when crucial differences arise between its members.

In summary, the well-functioning family has the ability: (a) to identify conflict; (b) to tolerate its occurrence rather than recoil from it; (c) to permit its members to express views whether of resentment, anger, or criticism; (d) to encourage expression of corresponding emotion thus avoiding any incongruence (as for example in the situation of the double bind in which a message is inherently baffling because it is accompanied by a parallel, contradictory communication); and (e) to apply skills of negotiation wherever possible in the spirit of seeking compromise and resolving differences.

This last point brings us back to adaptive family communication patterns overall. Accepting the notion that communication reflects relationships, a desideratum for its effectiveness in the family is the mutual respect and understanding members have for one another. If this is the case, they are able to express themselves authentically and freely; moreover, differences of opinion are permitted. Yet a caveat applies, the mutual respect for privacy. Family members may find it necessary to maintain certain, impermeable boundaries (in systemic terms) within which they are able to preserve their innermost thoughts, feelings, and fantasies. An adolescent child for example may not wish to divulge her most personal matters to the family but prefer an intimate peer as confidante while her parents would probably concur that feelings about their sexual relationship are not relevant to their children.

Finally, the well-functioning family is aware of the salience of patterns of communication in that they are equipped to 'communicate about communication' (meta-communication is the relevant term here). They are able to discern between digital and analogic forms, detect incongruence, and identify contradictions or inconsistencies. In so doing, they are sensitive to their inner experience and, concurrently, can act empathically with those with whom they communicate.

THE FAMILY'S LINKS WITH ITS SURROUNDING CULTURE

Although, as Ackerman (1958) suggests, 'the family is the basic unit of growth and experience, fulfilment or failure', open channels between it and the surrounding 'culture' are necessary for its effective functioning. By culture we refer to an assortment of groups and influences (see Chapter 3). Groups include the extended families of husband and wife (in the case of the Western nuclear family), children's peers, work colleagues, the church, the neighbourhood, and friends. Less tangible cultural influences cover such features as ethnicity, poverty, racism, sexism, unemployment, and social services. These influences can be either detrimental or beneficial. Consider ethnicity; McGoldrick and her colleagues (1982) assert that it:

'. . . remains a vital force . . ., a major form of group identification and a major determinant of our family patterns and belief system'. Contrariwise, racism invariably exerts a harmful effect (see, for example, Littlewood and Lipsedge 1982; and Steere and Dowdall 1990)

Whatever the nature of the cultural environment, it is advantageous for the family to maintain an open system by having a distinct but permeable boundary between itself and the external world. This is essential for facilitating bidirectional traffic. On the one hand, members – singly or as part of sub-systems (for example, the children) or as a complete group – can confidently traverse the boundary, so benefiting from creative contact with the wider system including groups like peers, work colleagues, and school teachers. On the other hand, the family is actively receptive to appropriate environmental stimuli and therefore exposed to novel, challenging experiences with a possible enhancement of its functioning and development.

Reiss (1981) has shown from his laboratory research that such families are 'environment-sensitive' (see Chapter 2); they are oriented externally and interested as a group with the world. For example, in performing a shared task they observe as many cues as possible from each other and from the external environment. They also exhibit an appropriate balance between preserving personal boundaries and maintaining closeness. These families enjoy a confidence in being able to understand and master their social world; they can effectively explore and comprehend its nuances. They can also 'make conspicuous contributions to it as well as effectively express their own needs'. We shall return to Reiss's model in Chapter 2.

THE FAMILY LIFE CYCLE

We have alluded at several points to the idea of the family having a life cycle of its own. The concept is so central in the context of the well-functioning family (and indeed the malfunctioning too) that we now focus on it more explicitly.

Following Erik Erikson's (1968) pioneering contribution on the notion of the individual person passing through a series of stages, components of a cyclical pattern commencing in infancy and ending in senescence, the corollary of a comparable cycle in the life of the family is an obvious, conceptual sequel.

The sequel in fact was taken up by a sociologist, Evelyn Duvall (1977), who initially proposed a typical family life cycle. She identified eight stages covering the movement from the newly married couple through to the ageing family with parents retired and facing widowhood and their own deaths. In between came the new-child bearing stage, the family with preschool children, the family with school-going children, the family with adolescents, the family launching the late adolescent/young adult, and the 'empty nest' phase in which the offspring have departed from home. In short, we can talk about expanding and contracting families, dependent on the status of the children.

Several alternative models of the family life cycle have been advanced since Duvall's pioneering effort, most commonly by family therapists. The models deviate minimally from Duvall's schema, involving preferences in dividing the cycle in particular ways – from four to 24 stages! Nonetheless, all variations embody a similar sequence. This varied approach is arbitrary; more important is the appreciation of a series of developmental steps during each of which the family is faced with predictable but as yet unexperienced changes (except vicariously).

These changes in turn require major shifts in the family (so called second-order changes, in Watzlawick *et al.*'s terms (1974) in respect of its overall functioning, roles occupied by members, and rules formulated to serve as guidelines. The most illuminating approach is that by Carter and McGoldrick (1980). Their edited volume entitled '*The family life cycle: A framework for family therapy*' is mandatory reading for those who wish to learn more about all facets of the subject including separated, divorced, and single-parent families, women in families, and the multi-problem, poor family. There are also comprehensive chapters on the specific stages of the family cycle.

Carter and McGoldrick themselves offer a six-stage schema in which each stage is associated with the chief emotional process of the transition involved, and with systemic changes necessary for continuing development:

1. In the first stage the unattached young adult has to develop a sense of separation from his or her parents with consequent evolution of a work role and of intimate peer relationships.

2. The second stage involves the newly married couple who must commit themselves to a new marital partnership based on mutual interdependence, and modify their relationships with the two extended families.

3. The family with young children constitutes the third stage; the partners now have to alter their marital relationship in order to accommodate children, assume parenting roles, and again reshape relationships with the extended families, especially the newly established grandparents.

4. The fourth stage, the family with adolescent children, entails the parents granting permission to the adolescent to establish a more autonomous

self while, at the same time, re-evaluating their own 'mid-life' issues such as marriage and career.

5. Launching the children is the principal challenge of the fifth stage. Several exits from the family now begin, requiring a shift in the relationship between parents and grown children to an adult–adult form. Additionally, the parents have to re-examine their own relationship as well as grapple with possible illness and death of their own parents.

6. The family in later life, the final stage, requires the acceptance of a change in generational roles. The original parents (grandparents now) must seek new roles within the family and beyond; they must also be offered appropriate support by the now established middle generation as they deal with death of spouse, siblings and other peers, and anticipation of their own death.

One purpose of spelling out the schema in detail is to make the point, albeit indirectly, that each stage, and equally important, each transition between stages, is accompanied inevitably by pressures which exert major demands on all family members. To quote Carter and McGoldrick:

Our hypothesis is that there are emotional tasks to be fulfilled by the family system at each phase of its life cycle, requiring a change in status of family members, and that there is a *complex emotional process* [our italics] involved in making the transition from phase to phase.

Putting it another way, Haley (1973) comments: 'Whatever the stage of family life, the transition to the next stage is a crucial step in the development of a person and his family'.

The features of the well-functioning family discussed earlier are particularly pertinent at these times. Aspects like adaptability to change, flexibility, cohesiveness, tolerance of conflict, and clarity of communication are crucial to the effective negotiation of what is in reality a life-long, highly demanding *rite de passage*.

This is made all the more demanding by virtue of the fact that in addition to the aforementioned predictable sequence a series of 'accidental' life events including chronic illness and disability, migration, disaster, war, and premature death may occur. These events may obviously constitute a source of profound stress and call on the family to harness all its resources in order to deal with them effectively.

Given the inevitability of this variegated set of pressures over an extended period, coupled with the notion of development along the life cycle being epigenetic in nature (that is, dealing with the next stage in the sequence is dependent on the effects of negotiation of the preceding stage), it comes as no surprise that a particular stage may not be fully dealt with. Two family therapists, Barnhill and Longo (1978), have cogently pointed this out and incorporated the psychodynamic concepts of fixation and regression to make for a more complete understanding of the family's progression through its life cycle. They argue this way:

Just as in the case of the individual, it is possible to hypothesize that families pass through and resolve the conflicts of each stage with varying degrees of success. Since 100 per cent success at resolving the conflicts is rare, it can be assumed that there will be some partial fixation on unresolved issues at one or several of the life cycle stages.

Since additional changes are necessary as the family cycle continues, unresolved issues, the authors suggest, are 'sealed over', leaving foci of vulnerability. Moreover, in later stressful circumstances, the family might conceivably resort to previous modes of functioning. This leaves the family wrestling with two sets of problems – one derived from the current stress, the other from unresolved difficulties of a previous stage. Although Barnhill and Longo raise these issues in a clinical context, it seems likely that a similar scenario prevails in non-clinical families.

Another facet of this notion of partial dislocation of progression in the cycle relates to the family undergoing substantial structural change, the most obvious and most common being that of separation or divorce. Here, all manner of change may be required from a new courtship period for the divorcee and his or her new partner in the presence of children from one or two previous nuclear families to a widow's struggle on her own with the 'empty nest' phase. Aware of such possibilities, Carter and McGoldrick (1980; see Tables 1.2 and 1.3 in text) have usefully modified their basic family cycle for the intact family to cover effects on the cycle when applied to divorce, single parent families, and reconstituted families.

Notwithstanding the customary focus on the Western, nuclear family, the family cycle in its basic dimensions is relevant to all family groups. As Minuchin and Fishman (1981) posit: '. . . whatever the circumstances, the basic flow remains: a family has to go through certain stages of growth and ageing. It must cope with periods of crisis and transition'.

Having spelled out the six areas of family life deemed relevant in determining the well-functioning family, and briefly considering the family life cycle, we look more closely in the next chapter at the main models of family function that have been elaborated and systemically studied.

REFERENCES

Ackerman, N. (1958). *The psychodynamics of family life*. Basic Books, New York.
Barnhill, L.R. and Longo, D. (1978). Fixation and regression in the family life cycle. *Family Process*, **17**, 469–78.
Bateson, G. (1972). *Steps to an ecology of mind*. Ballantine, New York.

Bertalanffy, L. von. (1968). *General systems theory*. Braziller, New York.

Boszormenyi-Nagy, I., Grunebaum, J., and Ulrich, D. (1991). Contextual therapy. In *Handbook of family therapy*, Vol. II (ed. A. Gurman and D. Kniskern). Brunner/Mazel, New York.

Bowen, M. (1961). The family as a unit of study and treatment. *American Journal of Orthopsychiatry*, **31**, 40–60.

Byng-Hall, J. (1988). Scripts and legends in families and family therapy. *Family Process*, **27**, 167–79.

Carter, E. and McGoldrick, M. (ed.) (1980). *The family life cycle: a framework for family therapy*. Gardiner Press, New York.

Dunn, J. and Mann, P. (1985). Becoming a family member: Family conflict and the development of social understanding in the second year. *Child Development*, **56**, 480–92.

Duvall, E. (1977). *Family development*. Lippincott, Chicago.

Eisler, I., Dare, C., and Szmukler, G. (1988). What's happened to family interaction research? An historical account and a family systems viewpoint. *Journal of Marital and Family Therapy*, **14**, 45–65.

Epstein, N.B., Bishop, D.S., and Baldwin, L.M. (1982). McMaster model of family functioning: A view of the normal family. In *Normal family processes* (ed. F. Walsh). Guilford, New York.

Erikson, E. (1968). *Identity, youth and crisis*. Norton, New York.

Falloon, I., Boyd, J., and McGill, C. (1984). *Family care of schizophrenia: a problem-solving approach to the treatment of mental illness*. Guilford, New York.

Haley, J. (1963). *Strategies of psychotherapy*. Grune and Stratton, New York.

Haley, J. (1973). *Uncommon therapy*. Norton, New York.

Jackson, D. (1965a). The study of the family. *Family Process*, **4**, 1–20.

Jackson, D. (1965b). Family roles: The marital quid pro quo. *Archives of General Psychiatry*, **12**, 589–94.

Jackson, D. (1977). The myth of normality. In *The interactional view* (ed. P. Watzlawick and J. Weakland). Norton, New York.

Janosik, E. and Green, E. (1992). *Family life*. Jones and Bartlett, Boston.

Jenkins, H. (1989). Precipitating crisis in families: patterns that connect. *Journal of Family Therapy*, **11**, 99–109.

Kohlberg, L. (1981). *Essays on moral development: Vol. 1, The philosophy of moral development: moral stages and the idea of justice*. Harper and Row, New York.

Luepnitz, D. (1988). *The family interpreted: feminist theory in clinical practice*. Basic Books, New York.

Littlewood, R. and Lipsedge, M. (1982). *Aliens and alienists: ethnic minorities and psychiatry*. Penguin, Harmondsworth.

McGoldrick, M., Pearce, J., and Giordano, J. (1982). (ed.) *Ethnicity and family therapy*. Guilford, New York.

Minuchin, S. (1974). *Families and family therapy*. Harvard University Press, Cambridge, Mass.

Minuchin, S. and Fishman, H.C. (1981). *Family therapy techniques*. Harvard University Press, Cambridge, Mass.

National bioethics consultative committee (1991). *Reproductive technology counseling*. Report of March, 1991.

Olson, D., Sprenkle, D.H., and Russell, C.S. (1979). Circumplex Model of marital and family systems: I. Cohesive and adaptability dimensions, family types, and clinical application. *Family Process*, **18**, 3–28.

Piaget, J. (1977). *The essential Piaget*. (ed. H. Gruber and J. Voneche) Basic Books, New York.

Reiss, D. (1981). *The family's construction of reality*. Harvard University Press, Cambridge, Mass.

Satir, V. (1967). *Conjoint family therapy*. Science and Behaviour Books, Palo Alto.

Shorter, E. (1975). *The making of the modern family*. Basic Books, New York.

Simon, F.B., Stierlin, H., and Wynne, L. (1985). *The language of family therapy: A systemic vocabulary and sourcebook*. Family Process, New York.

Steere, J. and Dowdall, T. (1990). On being ethical in unethical places: The dilemmas of South African clinical psychologists. *Hastings Center Report*, **20**, 11–15.

Watzlawick, P., Beavin, J., and Jackson, D. (1967). *Pragmatics of human communication*. Norton, New York.

Watzlawick, P., Weakland, J., and Fisch, R. (1974). *Change: principles of problem formation and problem resolution*. Norton, New York.

Weakland, J., Fisch, R., Watzlawick, P., and Bodin, A. (1974). Brief therapy: Focused problem resolution. *Family Process*, **13**, 141–68.

Wynne, L.C., Jones, J.E., and Al-Khayyal, M. (1982). Healthy family communication patterns: observations in families 'at risk' for psychopathology. In *Normal family process* (ed. F. Walsh). Guilford, New York.

2

Models of family function and their measurement

Chapter 1 had as its main focus an account of the criteria of the well-functioning family. We commented at several points upon the contribution of systematic researchers; in this chapter we look in detail at their work. We consider the main groups in the field, all as it happens from North America, where family researchers have been most active. The models to be covered are the McMaster, Family Environmental, Circumplex, Family Paradigm, and Beavers System. The chapter ends with a brief discussion of the potential integration of these models. Table 2.1 shows, in summary form, the chief ingredients of all the models dealt with in the chapter.

MODELS OF FAMILY FUNCTION BASED ON SELF-REPORT

The McMaster model

Nathan Epstein and his colleagues have influenced the study of family function since the 1970s. Their work in fact commenced in the previous decade when Westley and Epstein (1969) studied families in which there was a student member. The first phase saw a detailed study of nine families; of the 59 families comprising the sample in the second phase, 10 with the most disturbed students were compared with 10 containing the 'healthiest' students, on such variables as status, roles, power, and psychodynamics.

Two major findings emerged: interactional factors were more relevant than intrapersonal ones in affecting family members' behaviour; and the students' psychological condition correlated closely with the quality of their parents' relationships. Not too surprisingly, when the marital relationship was stable, the students enjoyed good psychological health. The research was later transferred to McMaster University (hence the model's name), and subsequently, in 1982, to Brown University. The culmination of this painstaking work was the McMaster model of family function which we will now consider.

Epstein *et al.* (1982) begin with the notion that the family faces three

Table 2.1 Chief ingredients of models of family function covered in this chapter

McMaster	Family Environment	Circumplex	Paradigm	Beavers	
				Competence	*Style*
Problem-solving	*Relationship*	Cohesion	Configuration	Structure:	Dependency needs
Communication	Cohesion	Adaptability	Co-ordination	overt power parental coalition closeness	Adult conflict
Family roles	Expressiveness		Closure		Proximity
Affective responsiveness	Conflict			Mythology	Social presentation
Affective involvement	*Personal growth*			Goal-directed negotiation	Expression of closeness
Behaviour control	Independence				Assertive and aggressive qualities
	Achievement orientation			Autonomy:	Expression of feelings
	Intellectual-cultural			clarity of expression responsibility permeability	
	Active-recreational				
	Moral-religious			Family affect: range of feelings mood and tone unresolvable conflict empathy	
	System maintenance				
	Organization				
	Control				

Family Assessment Device (FAD)	Family Environment Scale (FES)	Family Adaptability and Cohesion Evaluation Scale (FACES)	Observer ratings	Observer ratings	Observer ratings
Self-report	Self-report	Self-report			

sets of tasks throughout its life cycle: instrumental (or basic), covering such needs as food, accommodation, and financial security; developmental – predictable problems and challenges inherent in the family life cycle; and crisis or hazardous – unexpected life events such as illness, accidents, and migration. Furthermore, the model is based on general systems theory: the family parts are inter-connected; no part can be understood outside of its context; and the family's structure and function, including its interactional patterns, are important determinants of individual members' behaviour.

With these points in mind, we can readily appreciate the multidimensional nature of the model. The six dimensions covered in the Family Assessment Device (FAD), the scale devised to measure the McMaster model, are:

1. *Problem solving* – the family's ability to deal with the three types of tasks without losing effectiveness as a system. As mentioned in Chapter 1, problem solving requires a number of sequential steps – problem identification, communication about the problem with relevant people, devising alternative remedies, selecting one of them, taking necessary action, monitoring the action and appraising its level of success.

2. *Communication* – in the well-functioning family, exchange of information and corresponding emotion between members has two main features: clarity – the message is unambiguous thus diminishing the chance of its being misconstrued; and directness – the message is targeted at those to whom it is intended. Non-verbal behaviour is less crucial than verbal communication but congruence between them makes for superior clarity and directness.

3. *Family roles* – these are defined as consistent patterns of behaviour by which family functions are executed. Roles are filled in the well-functioning family without any member becoming unduly burdened. Roles are fairly allocated, responsibility is clearly defined and monitored, and room is left for flexible reassignment. Roles are required for, *inter alia*, nurturance, support, personal development of members, leadership, decision-making, and control.

4. *Affective responsiveness* – this concerns the family's capacity to react to circumstances with appropriate feeling, in both quality and degree. In a well-functioning family, members can express themselves freely without emergent feelings becoming excessively intense or prolonged.

5. *Affective involvement* – this reflects the family's display of interest in fellow members and may range, on a continuous dimension, from total absence to symbiotic ties. Empathic involvement is typical of the

well-functioning family in which accurate appreciation of the needs of others occurs.

6. *Behaviour control* – this final feature concerns the family's capacity to handle three types of situation: danger, interpersonal relating both within and beyond the family, and the meeting of members' psychological needs. Behaviour control may be rigid, flexible, *laissez-faire*, or chaotic. A flexible approach typifies the well-functioning family.

The dimensions are measured by each member rating his level of agreement on each of 60 items. For instance, 'We are reluctant to show our affection for each other' is one of the 10 items covering affective responsiveness while 'Family tasks don't get spread around enough' concerns family roles. An overall level of family function is also derived from the FAD.

The instrument has been psychometrically appraised and found to have reasonable levels of internal consistency, test–retest reliability, and discriminant validity, that is, differentiating between families rated by experienced clinicians as functional or dysfunctional. Like all self-report measures, the FAD is difficult to interpret when results from several members are poorly correlated. Does this point to differing perceptions of the same family phenomena or do some respondents achieve greater accuracy than others on what is the 'truth of the matter'? Obviously, these questions are difficult to clarify but the approach, adopted by Epstein and his team, of averaging scores of all members is not especially illuminating; on the contrary, it may obscure genuine differences of perception or constitute a distortion of the family's 'general' viewpoint through extreme scoring in cases. We shall return to this important issue at the end of the chapter.

A clinical rating scale has also been devised by the Epstein group (1983) which could obviate the self-report difficulty but it is probably more suited to the assessment of dysfunctional families in the clinical situation.

The 12-item 'general function' sub-scale of the FAD has been usefully examined in a community sample of 2000 Canadian families. A parent, usually the mother, was also interviewed about family difficulties. Reliability, internal consistency, and split-half coefficients were all satisfactory. Especially important was the finding on construct validity, that is, there were close relationships between scores on the 'general function' scale and several family variables included in the Ontario study (Byles *et al.* 1988).

Another model of family functioning to have emerged from Canada is the *Process Model* with its corresponding Family Assessment Measure (FAM). In fact the FAM is derived from the same source as the FAD; given their similarity, we will deal only briefly with the former. Its authors, Skinner, Steinhauer, and Santa-Barbara (1983), sought a model that reflected an integration of systems, psychoanalytic, attachment, social learning, and crisis theories of development and psychopathology. This approach, they argued, would permit an emphasis on *both* intrapsychic and interpersonal factors as well as consider the interface between the family and larger social systems.

A departure from the McMaster model (and having something in common with the Family Environment Scale to which we shortly turn) is the addition of the family's values as influencing several other dimensions of the Process Model. The Canadian group (Steinhauer *et al.* 1984) avers that a family's values are crucially important for its function because they 'define the context within which all other dimensions of the model operate . . . [and have] pervasive direct and indirect influences on all aspects of family functioning'.

The Family Environment approach

Another pioneering approach to family function but one originally geared to non-clinical families was developed in the 1970s by Moos (Moos and Moos 1981), a psychologist working in the framework of social ecology and well known for his measurement of various 'social environments'.

Unlike the previous two groups, Moos did not set out with a specific model of family function in mind but on the premise that all social groups have accurately measurable features. Thus, he assembled a large pool of potential items reflecting family function by interviewing families and scrutinizing existing 'social climate' scales. After the resultant form had been administered to 1000 subjects, 90 items were extracted and 10 sets of items identified through a factor analysis (a brief version contains 40 items). The 10 sub-scales making up the Family Environmental Scale (FES) are grouped into three categories as follows:

(A) *Relationship dimensions* – three sub-scales deal with interrelationships:

1. Cohesion – the degree to which members are committed to one another and are mutually helpful and supportive (for example, 'There is a feeling of togetherness in our family').

2. Expressiveness – the degree to which members express their feelings directly (for example, 'Family members often keep their feelings to themselves').

3. Conflict – the degree to which members express anger and aggression and are engaged in conflict (for example, 'Family members sometimes get so angry they throw things').

(B) *Personal growth dimensions* – five sub-scales are concerned with aspects of personal growth:

1. Independence – the degree to which members assert themselves, are self-sufficient and make their own decisions (for example, 'There is one family member who makes most of the decisions').

2. Achievement orientation – the degree to which certain activities are construed in an achievement-oriented or competitive framework (for example, 'We believe in competition and "may the best man win"').

3. Intellectual-cultural orientation – the degree to which the family is interested in political, social, intellectual, and cultural activities (for example, 'We often talk about political and social problems').

4. Active-recreational orientation – the degree to which members participate in social and recreational activities (for example, 'We are not that interested in cultural activities').

5. Moral-religious emphasis – the degree to which ethical and religious issues and values are emphasized within the family (for example, 'Family members attend church or Sunday-school fairly often').

(C) *System maintenance dimension* – two sub-scales cover organizational aspects:

1. Organization – the degree of importance attached to organization and structure in planning activities and assigning responsibility (for example, 'Activities in our family are carefully planned').

2. Control – the degree to which explicit rules and procedures are used to organize the family's life (for example, 'There are set ways of doing things at home').

Scores on the 10 sub-scales are obtained by respondents indicating whether a statement is true or false about their family. Responses reflect what Moos has termed the family 'climate'; another phrase used is the family's 'social-environmental characteristics'. The FES is dependent on the respondents' perceptions of the family along the 10 dimensions. Since various members may see family life differently, Moos introduced the concept of a family incongruence score, that is, level of agreement between members. This procedure enables the researcher to compare concordance in different sorts of families, for example, between well-functioning and dysfunctional families.

In addition to the regular form, the FES has also been used to elicit what the respondent expects the family climate to be like, the so-called *expectation* form. The *ideal* form calls on respondents to reflect on the family they would ideally wish to be part of. A highly sensitive measure obtained from the FES, the *family relations index*, is a summation of two relationship dimensions, cohesion and expressiveness, minus conflict, the third relationship dimension.

Scoring is straightforward. Raw scores on the sub-scales are converted into standard scores which then permit profiles to be portrayed. Unlike the FAD and FAM, FES scoring applies the notion of the bell-shaped curve, that is, families at either end of the continuum are 'statistically' different. The snag however is the inability to label families as dysfunctional.

The FES was subject to extensive psychometric examination early on in its existence and was shown by Moos (Moos and Moos 1981) to have satisfactory internal consistency and test-retest reliability. The sub-scales also had low to moderate intercorrelations indicating their separateness from one another. Discriminant validity was established by comparing distressed with healthy families.

Rather salutary however was the publication in 1990, that is, several years after the FES's widespread use, of criticism of its psychometric qualities by Roosa and Beals (1990). The FES has indeed been deployed in dozens of studies over the past decade. Especially prominent has been a series of investigations on the outcome of therapy of alcoholism (see, for example, Moos *et al.* 1979). Other topics covered include issues in family therapy (Fuhr *et al.* 1981); its use to identify types of family environments (Moos and Moos 1976); and in health research (for example, Marteau *et al.* 1987). Roosa and Beals (1990) found internal consistencies of the five sub-scales they examined below the acceptable level necessary for practical or research use, and certainly much lower than those obtained by Moos. Furthermore, ratings performed by students cast doubts on the validity of the sub-scales. The authors concluded that: 'These results raise serious concerns about future use of the FES. Researchers should be cautious in choosing to use the FES until more is known about its reliability across samples and its application to their specific sample'.

Moos (1990) fought back, claiming that the FES is indeed reliable in terms of internal consistency and does have construct, concurrent, and predictive validity. He also asserted that in order to advance family assessment measures, researchers needed to apply conceptual as well as psychometric criteria and not rely excessively on internal consistency, reliability, and factor analytic approaches to the development and validation of scales.

The technical arguments of both parties can be pursued further by careful perusal of the relevant papers. For our purposes, we should take note of the conceptual and methodological hurdles which stand in the researcher's (and clinician's) way in their endeavour to apply family self-report scales, and when the target of assessment is a group rather than an individual.

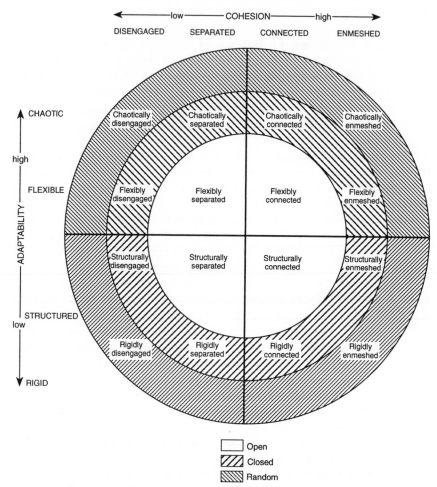

Fig. 2.1 The circumplex Model: sixteen possible types of marital and family systems (Olson 1986).

The Circumplex model

By contrast with the McMaster model, the Family Process model and the Family Environment concept, the Circumplex model, developed by Olson and his colleagues (Olson *et al.* 1983), contains only two dimensions, cohesion and adaptability. According to their hypothesis, both are mediated by communication. On the other hand, no less than 16 family types are generated by the instrument devised to measure the two dimensions.

Cohesion and adaptability were originally selected by Olson as fundamental as a result of '. . . inductive clustering rather than from an empirical clustering based on factor analysis [as with the FES]' (Olson *et al.* 1979). The choice was informed by GST to which Olson pays tribute for its appropriateness in studying the family.

Cohesion, reflected in at least 40 concepts used previously in the social sciences literature, and defined as the 'emotional bonding members have with one another and the degree of individual autonomy a person experiences in the family system', was regarded by the researchers as especially salient. In the Circumplex model, its range extends from an extreme of high cohesiveness or enmeshment, where bonding is excessive, individual autonomy limited, and over-identification with the family prominent to an opposite extreme of disengagement 'characterised by low bonding and high autonomy from the family'.

Adaptability, the other dimension, has been defined as the: 'ability of a marital/family system to change its power structure, role relationships and relationship rules in response to situational and developmental stress [inherent in the family life cycle]'. Related concepts include negotiation style, feedback – both positive and negative, flexibility, and preparedness to adjust. The range here is from an extreme of rigidity which prevents the family from experiencing any flexibility in handling a variety of situations to chaos, viewed by the research group as an excessive form of adaptability (this latter notion may seem a contradiction but more about this anon).

Given the posited, orthogonal relationship between the two dimensions and their potential range, the model yields 16 family types. As can be seen in Fig. 2.1, the four types in the central area – flexibly separated, flexibly connected, structurally separated, and structurally connected – comprise the most functional systems whereas those in the outer circle – chaotically disengaged, chaotically enmeshed, rigidly disengaged, and rigidly enmeshed – are associated with dysfunction.

Moreover, Olson hypothesizes that the balanced family types are most conducive to effective family function and to optimal individual development, that is, a curvilinear relationship exists between family function and each of the two dimensions.

Both cohesion and adaptability are measured on a 20-item self-report questionnaire, the Family Adaptability and Cohesion Evaluation Scales, or FACES III (its third version) (Olson 1986). The questionnaire elicits the respondent's perception of his family. Scores on the two dimensions are plotted onto the model as in Fig. 2.1, thus portraying the type of family system perceived. As with the FES, an ideal form has been devised which seeks the picture the respondent would like to envisage of his family. The difference between the two scores yields a measure of the degree of satisfaction a person has with his family; obviously, the greater the gap the greater the dissatisfaction.

Items in FACES III illustrative of cohesion are: 'Family members ask each other for help', 'Family members feel very close to each other', and 'We can easily think of things to do together as a family'. Reflecting adaptability are: 'Rules change in our family', 'Our family changes its way of handling tasks', and 'We shift household responsibilities from one person to another'.

A clinical rating scale is also available which permits an outsider's perspective.

Olson and his team have reported on the psychometric properties of their scales using a large, normative sample of 2500 adults. Test-retest reliability was good after a four to five week interval. Similarly, internal consistency was satisfactory. The authors have also lodged a claim for face and content validity but not for concurrent validity. Good evidence is also claimed for FACES III in distinguishing between different types of family groups.

As mentioned earlier, Olson's assumptions that balanced family types are more functional than extreme types is questionable. This has become controversial following initial studies which did in fact demonstrate a curvilinear relationship. An investigation of its refutation is exemplified in the work of Anderson and Gavazzi (1990). They found a consistent pattern in a clinical sample of 110 adults, that most patients had extreme scores on either cohesion or adaptability but examination of the relationship between the two dimensions and the respondents' perceptions of their personal, marital, and family function pointed to a linear pattern. Thus, the question arises: is the model itself or are the instruments derived from it problematic in permitting the model to be adequately tested? Anderson and Gavazzi hesitate to unravel this issue because of a lack of data. Instead, they conclude that the conceptual foundation of the model remains unproven and cannot be adequately studied until the reliability and validity of the questionnaire purporting to measure adaptability and cohesion are more soundly established.

In a much larger study, of 2440 families (Green *et al.* 1991), the researchers found no evidence of curvilinearity between either cohesion or adaptability and measures of personal and marital function. They concluded: 'The use of a comprehensive set of data analysis procedures in a sample with adequate size and optimal variability failed to confirm the predicted curvilinear relationship'.

Reference to the nature of the sample should remind us that not all well-functioning families share identical properties. Ironically, Olson (1986) himself makes this point when referring to cultural and ethnical diversity as a factor in determining family function (see Chapter 3). He even advances the hypothesis that:

'. . . if normative expectations of families support behavioural extremes on one or both of the dimensions, families will function well as long as all family members are satisfied with these expectations. In this way, the family serves as its own norm base'.

We have not explored this basic issue concerning the model out of a preoccupation with methodological matters (although we do regard them as important) but to make the following crucial point: when the model was first published in 1979, it was seized upon by researchers as a godsend. Literally hundreds of studies incorporated the model. Only latterly, through the publication of critical reports, has unease manifested about the shakiness of the assumed curvilinear relationship between the model's two parameters and family well-being. The message is clear. Those working in the family field whether as investigators or clinicians have need for models of family function and will clutch at those that appear coherent and relevant. But the possibility follows that they may do so without adequate reflection and without a firm assessment of the psychometric qualities of the instruments devised to test the model.

These comments also refer to the FES as we noted when discussing that instrument; it is also used in dozens of studies on the assumption that it is psychometrically adequate.

MODELS OF FAMILY FUNCTION BASED ON OBSERVATION

Having dealt with the chief models of family function based on self-report, we now focus on those which depend on ratings derived from observation of a family performing a standardized task. Other forms of observer ratings have been devised, especially in the clinical arena but given our theme of the well-functioning family, an account of these is beyond our remit (see, for example, the 'Standardized Clinical Family Interview', a research interview which highlights family interaction in the clinical setting (Kinston and Loader 1984) and (Kinston and Loader 1986)).

Two models have been pre-eminent since the mid-1970s – Reiss's Family Paradigm model and the Timberlawn or Beavers System model. We briefly consider them below.

Family Paradigm model

David Reiss's initial interest in families revolved around possible inter-actional disturbances in schizophrenic families. This was in the era following publication of the double-bind theory by Bateson *et al.* (1956). In studying these families, Reiss noted that more pertinent than their performance of a given task was their perception of the laboratory setting (a metaphor for the social setting) in which they were being assessed. His focus turned to the family's pattern of response to the novel circumstances, which he hypothesized paralleled their customary attitude towards the world in general.

He soon advanced the idea that a family's approach to problem solving derives from a shared world view (encompassing assumptions, expectations, and fantasies) and coined the term 'paradigm' to reflect this view (borrowing it from Thomas Kuhn).

The metaphorical use of Kuhn's concept of paradigm is helpful in aggregating these meanings. We now speak of the family paradigm as a central organiser of its shared constructs, sets, expectations and fantasies about its social world. Further, each family's transactions with its social world can be distinguished one from the other by the differences in their paradigms.

(Reiss 1981)

In research terms, the hypothesis followed that a family's approach to construing reality and the social world, manifested in the way it processed information and solved problems in the laboratory situation. Families' perceptions of a psychiatric ward, and of fellow families in multiple family therapy have also been used in Reiss's studies). By standardizing the task and related laboratory conditions, an observer was well placed to distinguish between various types of families. Reiss did concede however that the link between paradigm and overt behaviour was complex, providing for indirect observation only.

A description of the tasks set is not relevant here. Suffice to say, the family, usually a triad comprising two parents and an adolescent child, were asked to identify underlying patterns in sequences of circles, triangles, and squares or to sort cards into categories of their own making or to sort out a lattice 'puzzle' – with each member given a unique set of information and required to share it with other family members in order to complete the task (Jacob and Tennebaum 1988). Foci of interest for the observer were the family's approaches to gathering, interpreting, and exchanging information. In terms of family paradigm, measures included the degree of mutual influence, risk-taking, intermember control, and flexibility.

Scores on these factors were related to the three constructs which evolved out of the research:

(1) Configuration – the contribution that the family *as a group* makes to problem solving beyond their individual efforts (does the family enhance its problem solving ability compared to individual efforts?);

(2) Coordination – the inclination of the family to develop similar solutions; and

(3) Closure – the family's willingness to be open to new information and its corresponding capacity to tolerate delay in making decisions.

Since observers' ratings are objective, for example, time required to complete a task, reliability is high. The Reiss team also conducted a series of investigations to consider various types of validity. The laboratory procedure distinguished between 'normal' families and those with a schizophrenic or personality disordered member. The procedure has also enabled the prediction of a family's perception of other families. Moreover, configuration, coordination, and closure have been demonstrated to be independent of each other

Family typology in the Paradigm model
A typology of families has emerged from this research as summarized below:

1. The *environment-sensitive* family, representing the well-functioning family, is susceptible to as many cues as possible from each member and the environment. Members are capable, both as individuals and as a unit, of working towards an optimal solution to the problem being tackled; they are able to defer making decisions until they have obtained and evaluated pertinent information; all members accept that the agreed-upon solution is a function of both their own effort and of the family as a collaborative group. In terms of Reiss's three constructs, the environment-sensitive family scores highly on configuration, coordination, and closure (that is, remains open to new information and tolerates the experience of waiting).

2. *The interpersonal distance-sensitive family* is, by contrast, dysfunctional in dealing with the environment. Members do not accept assistance from each other; receiving information from one another is construed as weakness. Instead, they maintain distance in order to preserve their independence; and they exhibit disengagement, even pseudo-hostility. Reiss points out that in the extreme case, each member views the problem as part of his private world precluding collaboration. Closure may go to either extreme in the quest for independent action.

In terms of the constructs, the interpersonal distance-sensitive family scores poorly on coordination and configuration, and variably on closure.

3. *The consensus-sensitive family* finds it necessary to maintain a dogged, unified stance even though this disrupts the pursuit of an optimal solution. Members forgo autonomy and thus their own ideas in order to achieve consensus. Conformism is essential, dissent frowned upon. The world is seen as hazardous and unpredictable, beyond their understanding and control. Little reference is made to environmental cues which could facilitate

problem resolution: 'The family reaches its hastily forged consensus early in the task. If cues and information continue to be provided, the family distorts or oversimplifies them in order to justify its initial collective solution' (Reiss 1981, p. 70). This type of family is high on coordination and low on configuration and closure.

4. *The achievement-sensitive family* is staunchly set to make its own observations and creates its own relationship with the world. Members compete intensely with one another in pursuit of achievement. Since concordance is tantamount to surrender, they need to disagree with one another. In construct terms, the family is low on coordination but high on configuration (closure is less relevant).

We have provided a relatively detailed account of the model because of its solid empirical support and its interesting emphasis on the way the family perceives the outside world and its basic premises about it. Reiss demonstrates well how painstaking cumulative, systematic research (over 30 years) can enrich our understanding of family process by distinguishing between functional and dysfunctional. Application of constructs like configuration, coordination, and closure are also of utility in creating a typology of families in relation to function.

But there are snags with this meticulous methodology. Reiss himself raises some in his comprehensive 1981 research report. We would concur with him about one of his comments, namely the virtually exclusive emphasis in the model on the family's perception of the outside world. What about the family's conception of itself and particularly of intermember relationships? Surely, this intrafamilial dimension warrants attention.

Another matter concerns the premise that all family members have a similar view of the world. We would conclude, based on observation of both clinical and non-clinical families, that while this may be the case it is not at all uncommon for intergenerational and interpersonal differences and perhaps less marked, inter-sibling differences to occur.

Reiss's model, like the Circumplex model, may be too limited in its scope, failing to attend to other dimensions which underlie family function. As *part* of the 'story' the model is not only plausible but also convincing. We need to bear in mind however that it certainly requires to be complemented by other factors.

The Timberlawn or Beavers System model

A second, well-known measure of family function based on observation (although a self-report scale has been developed to complement the observational approach) derives from the Timberlawn or Beavers System model (Lewis *et al.* 1976, Beavers and Hampson 1990). Developed by Robert Beavers and his colleagues, the model and its associated assessment instruments have been at the forefront of family psychological research. The project's initial focus was on 'healthy' families. Subsequently, the measures have been deployed in the clinical arena.

The model classifies families along two axes, competence and style of interaction, and in so doing assumes that family function is best conceptualized as dimensional rather than categorical.

Family competence

Competence is a global term indicating a family's ability to perform its necessary tasks, for example, '. . . providing supporting support and nurturance, establishing effective generational boundaries and leadership, promoting the developmental separation and autonomy of its off-spring, negotiating conflict and communicating effectively'. It will be noted that several of these tasks are reminiscent of categories in the McMaster model.

In terms of measurement, competence comprises ten sub-scales; they are (each accompanied by a brief annotation):

(A) *Structure* (three dimensions)

1. Overt power – ranging from chaos where leadership is absent and the system is entropic (that is, in decline), through various degrees of dominance where leadership is rigid and authoritarian to egalitarianism in which the parents are equally important but not in every sphere of family life and co-leadership is flexibly assumed and relinquished.

2. Parental coalition – ranging from a sturdy parental coalition constituting consistent leadership at the most adaptive pole through weak co-leadership to its substitute by a parent-child coalition.

3. Closeness – blurred boundaries between members making for heteronomy typifies the dysfunctional pole, the range then extending through a sense of isolation among members to distinct boundaries which permit each member to be respected as separate and autonomous.

(B) *Mythology* (a single dimension)

At the functional pole the family is realistically aware of its function, this allowing for a high degree of concordance between its conceptions and those of the rater (sic). In the dysfunctional family, this concordance is lacking.

(C) *Goal-directed negotiation* (a single dimension)

This covers the family's ability to deal with problematic situations ranging from extremely inefficient to its direct opposite: '. . . what counts is that the family utilises its resources, personnel and time efficiently in negotiation of problem solutions'.

(D) *Autonomy* (three dimensions)

1. Clarity of expression – a rating of the degree of clarity with which feelings and thoughts are expressed. In competent families, clarity is optimal with members sanctioned to express themselves honestly and freely; the opposite prevails in incompetent families.

2. Responsibility – refers to the level to which members assume responsibility for their actions – past, present, and future. At one end of the scale responsibility is avoided and replaced by such behaviour as blaming or attacking others. At the opposite end responsibility is explicitly taken and regarded as a basic family rule.

3. Permeability – concerns the accessibility of members to one another's communications. The range extends from high receptiveness, a form of active listening, to a 'bland' disregard.

(E) *Family affect* (four dimensions)

1. Range of feelings – is concerned with the scope of expressed feelings not their intensity. Functional families show minimal expression whereas their healthy counterparts manifest a wide band.

2. Mood and tone – refers to the emotional quality of family interaction. Better functioning families exhibit emotional energy, spontaneity, humour, and optimism; poorer functioning families lack these features.

3. Unresolvable conflict – given that conflict is unavoidable in families, the question is how it is managed. At low levels of competence, negotiation of conflict is limited with grudges, confrontations, and seething evident. In the more competent family, unresolved conflict is negligible or absent.

4. Empathy – refers to the accurate understanding of members for one another. This is consistently prevalent in the competent family, grossly inappropriate or absent in the poorly functioning family.

Family style
Family style constitutes the second axis of family function in the model. Applying systems theory coupled with clinical observation, the Beavers

group has focused on one particular facet of behaviour, the family's inclination to interact either centripetally (CP), that is, to be inner-directed, regarding members as the chief providers of emotional support; or centrifugally (CF), that is, to turn to outside sources in the belief that they have more to offer than does the family (see also Stierlin 1972, Stierlin *et al.* 1973).

Style involves several components; in order to highlight differences in style, the following points out extreme features:

1. Dependency needs – CP families cling to one another whereas CF families ignore mutual needs for nurturance.

2. Adult conflict – CP families conceal conflicts in an effort to maintain harmony whereas CF families act directly and openly in the face of conflict.

3. Proximity – CF family members maintain distance between one another, finding intimacy disconcerting whereas CP members are close and achieve this comfortably.

4. Social presentation – CF families are not preoccupied with presenting a good face to the world whereas CP families strive to create a favourable impression.

5. Expression of closeness – CP family members emphasize their closeness to one another, avoiding any manifestation of dissent. CF families make it clear that they favour self-sufficiency.

6. Assertive and aggressive qualities – CF families to not hesitate to manifest assertiveness, regarding mutual challenge as worthy of encouragement; CP families are discomfited by any sign of aggression and make every effort to avoid expressions of anger.

7. Expression of positive and negative feelings (there is overlap here with the previous category) – CP families do not permit display of negative feelings such as anger and keep this sort of intense emotion to themselves. CF families on the other hand experience difficulty in expressing feelings of tenderness and affection.

Family typology in the Beavers model

Competence and style are regarded by Beavers and his colleagues as pivotal in understanding family function as well as in differentiating between 'healthy' and 'disturbed' families (see Beavers and Hampson, 1990). Figure 2.2 shows the schema whereby families are so categorized. We should however recall that both constructs are measured along a continuum. Although competence and style are not seen as orthogonal it will be noted that style *is* placed on the vertical and competence on the horizontal axes. This is a source of confusion in the model; the comment that '. . . [the] map of family functioning [is] based on theoretical, empirical and clinical work with families' fails to clarify the matter. Be it as it may, we note that healthy families, either 'adequate' or 'optimal', score highly on competence and in the mid-range on style. By contrast, 'borderline' and 'severely dysfunctional' families score poorly on competence and towards either end of the style continuum. This leaves

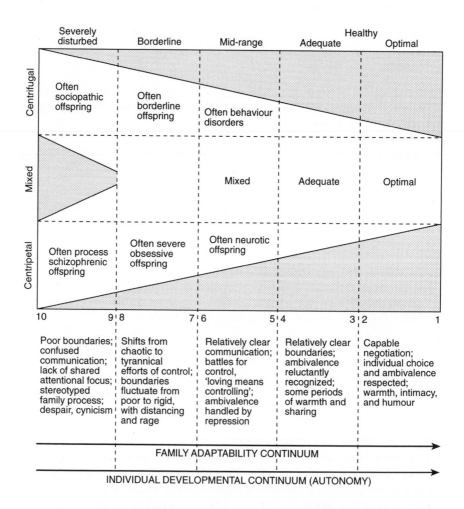

Fig. 2.2 Beavers Systems Model (Beavers and Hampson 1990).

a group of families labelled 'mid-range', which are placed midway between the extremes of competence and incompetence and display a style which is either a blend of centripetal and centrifugal or inclined to one or other but not predominantly so.

Ratings on the Beaver scales

We mentioned earlier that measurement of the model's constructs is based on observation. How is this accomplished? And how successful is the approach psychometrically?

Unlike the Paradigm model in which standardized tasks are administered, the Beavers model calls on family members to discuss, for 10 minutes, the changes they would like to see for themselves: 'Discuss together what you would like to see changed in your family'. This task is obviously suited to a family being assessed for or entering family therapy but is less apt for a 'research' family group. On the other hand, it would not be unreasonable to expect any family, clinical or otherwise, to tackle a task which requires its members to reflect on themselves as a group.

The task is done in the absence of a professional, therapist or researcher, but is videotaped or observed through a one-way screen. Teams of trained observers complete ratings on anchored scales (12 sub-scales for competence and seven for style), as well as two global scales (optimal to severely disturbed on competence, and centripetal to centrifugal on style).

Interrater reliability for the competence sub-scales and global scale is satisfactory – in the range of 0.72 to 0.89. Comparable figures for style are 0.62 to 0.83. Studies of validity are limited. In the original volume describing the model (Lewis *et al.* 1976), significant differences were found between scores on competence and on the global scale of 'healthy' and clinical families but the numbers involved were small (33 non-clinical and 70 clinical). The later volume by Beavers and Hampson (1990) barely mentions validity whether investigated by the Timberlawn group or others (although there are tests of concurrent validity on the Self-Report Family Inventory, that is, the self-report measure of the Beavers model). Similar data are presented for the style and global scales but in this case even the size of the samples is not provided. Thus, conspicuous omissions exist.

These points on psychometric aspects – that validity and, to a lesser extent, reliability have not been adequately tested, should induce a sense of caution. Moreover, other claims require clarification or further empirical testing. For example, the discussion by Beavers and Hampson (1990) on the link between competence and style, whether they are orthogonal or not, is unsatisfactory and certainly not buttressed by statistical evaluation. Overall, the criticism could be levelled that the model and its associated measurement are built on too many assumptions and on too few solid investigations.

COMPARING MODELS

Having considered several models and associated instruments purporting to measure dimensions of family function, we face the thorny questions: Why have so many models and measures been developed, and what accounts for the differences between them? The first question is easier to answer than the second. Research on the family is relatively recent; extensive gaps in knowledge have called out for investigation. The result has been the launch of several research groups, all setting out independently in the pursuit of a measurable model of family function. Dialogue between them only commenced in the 1980s after the second issue had crystallized, namely that their approaches, both conceptually and methodologically, differed substantially. Examples of these interchanges can be found particularly in the journal *Family Process* where much of the original and continuing research has been published (see, for example, Beavers and Voeller 1983; Sigafoos *et al.* 1985; Roosa and Beals 1990; and Moos 1990).

Cameron Lee (1988*a,b*) has placed these matters in perspective. Quoting Jay Haley who in 1962 had commented that the field then required a descriptive approach capable of classifying families and distinguishing between various types, Lee goes on to point out that during the ensuing quarter-century: 'Family researchers had worked diligently to construct theoretically and experimentally valid models to describe the complex phenomena of family life. Such models are invaluable to furthering and organising our knowledge of families' (p. 93–4).

Our account of this work in preceding pages leads us to concur with Lee and to surmise that inter-model differences may not be as extensive as on first inspection, having more to do with researchers' semantic predilections, particularly regarding the scope of the model (or to put it more concretely, the number of dimensions). This becomes apparent upon comparing the two chief approaches taken to exploring these differences – conceptual debate and empirical investigation. Let us consider representative examples of each.

An example of the *empirical approach* is the work of Bloom (1985). He undertook a study of four measures of family function – the FES, FACES II, FAM (all discussed in this chapter) and the Family Concept Q-Sort (a construct akin to an individual's self-concept and covering family members' attitudes and feelings about themselves) in order to identify, as he put it, 'a limited set of reliable concepts for describing families'.

Using cohorts of students, Bloom's strategy was to begin with one scale, maximize its psychometric features while at the same time reducing its length and then, in a series of cluster, factor, and correlational analyses, check whether the remaining scales covered dimensions of family function not previously identified. The exercise resulted in a new measure containing 15 dimensions, falling into one of three categories – relational, personal growth or values, and system maintenance (identical in fact to the FES categories). Most of these dimensions are familiar in one guise or another, similar to those of the source instruments, for example cohesiveness, enmeshment, disengagement, degree of conflict, expressiveness, intellectual-cultural orientation, sociability, democratic family style, and religious emphasis. Obviously, the number of dimensions exceeds those of any of the source questionnaires, resembling most closely the FES; but Bloom's scale also incorporates dimensions of the other three questionnaires included in his analysis.

The issue arising from this work is self-evident. A model of how the family works containing 15 dimensions is unwieldy to say the least and certainly not conducive to the task of determining *fundamental* features of family function and, the corollary, of distinguishing well-functioning from dysfunctional families. On the other hand, Bloom's study *does* suggest that after substantial research, the range of dimensions may be finite. Additional empirical study may indeed demonstrate that the number of dimensions can be further reduced.

Another relevant aspect is the procedure applied to measuring family function. Methodological differences could conceivably account for the lack of relationships encountered between models. For instance, in a comparison between the Circumplex and Paradigm models, no significant correlations (except one, in fathers) were found between the two dimensions of the former and the three dimensions of the latter (Sigafoos *et al.* 1985). The authors hypothesized that this could have been a function of the mode of measurement. In the Paradigm model (as discussed on page 32), the association between observer and observed is salient and in keeping with the nature of a model stressing the family's links with the wider, social environment. This is bound to influence the family's response to the investigator. In the Circumplex model, by contrast, the measure essentially deals with the family itself and therefore should not be affected by the interpersonal context of the research setting. A related assumption is made by Olson (1985) that congruence across models is only likely where similar methodology is applied.

Creative debate between proponents of various models is another pathway to the refinement of knowledge on family function. We mention 'creative' since the debate has not always been of that ilk. We can illustrate this by considering two important contributions to the literature. Lee's synthesis of the Circumplex and Beavers models serves as an example of a useful endeavour (Lee 1988*a*). Focusing particularly on adaptability, he contends that theoretical overlap between the two models is concealed by ambiguity in the Circumplex model. He proceeds to unravel interpretations of adaptability inherent in the models, the details of which need not concern us here. What follows is an innovative attempt to achieve a synthesis. Lee concludes with a sensible recommendation: '. . . the continuing dialogue between various models of the family, and the potential for their synthesis should play a central role'; elsewhere he comments: 'It is not a synthesis of models or of instruments but a synthesis of knowledge that I am suggesting. The fractionation of research programmes increases our knowledge of the family in a largely piecemeal fashion. A more integrated understanding will require a different kind of effort' (Lee 1988*b*).

The Beavers group (Hampson *et al.* 1988) tended to ignore Lee's recommendation, adopting a defensive stance around its model and a critical one towards 'rivals'. The following quote is typical:

[Lee's] conclusions regarding integration of the two models are unwarranted. When the conceptual and empirical difficulties of Olson's [Circumplex] model are adequately revealed, the combination of the models would not be a worthwhile effort. This is because the Beavers system model is a complete clinical, theoretical and empirical model unto itself, with demonstrated congruence between empirical and theoretical levels'.

The Olson group has been inclined to more flexibility, especially concerning the repeated finding that the associations between family function and cohesion and between family function and adaptability are linear rather that curvilinear, as claimed in the Circumplex model. Commenting on the finding of linearity by Green and his colleagues (1991), in a particularly good piece of research, Olson (1991) conceded that problems did exist in the conceptualization of his model and its related measure as well as in the scoring procedure. Moreover, he announced his planned collaboration with Green (essentially one of his critics) in an attempt to produce a new version of FACES, which he hoped would deal with previous theoretical and methodological limitations.

CONCLUSION

This chapter could easily be expanded into a book of its own, so assiduous have family researchers been in recent years. Indeed, two volumes published in 1989 reveal how much progress has been accomplished in family measurement specifically. The interested reader is recommended to read them (Grotevent and Carlson 1989; Touliatos *et al.* 1989). For our purposes, we have seen the nature and scope of the research conducted on devising and measuring models of how the family works. We may conclude that the results achieved hitherto have provided useful and illuminating insights into family function including pointers to the differentiation between dysfunctional and well-functioning families. We are thus better placed to appreciate the relevance of family features in families which contain a psychiatrically ill member.

In the next chapter we deal with some of the factors covered in this chapter and the previous one, in our account of family assessment in clinical practice. Our focus shifts sharply to the practical dimension but the aforementioned conceptual and empirical material should be kept in mind.

REFERENCES

Anderson, S.A. and Gavazzi, S.M. (1990). A test of the Olson Circumplex Model: Examining its curvilinear assumption and the presence of extreme types. *Family Process*, **29**, 309–24.

Bateson, G., Jackson, D.D., Haley, J., and Weakland, J. (1956). Towards a theory of schizophrenia. *Behavioural Science*, **1**, 251–64.

Beavers, W.R. and Hampson, R.B. (1990). *Successful families: assessment and intervention*. Norton, New York.

Beavers, W.R. and Voeller, M.N. (1983). Family models: Comparing and contrasting the Olson Circumplex Model with the Beavers System Model. *Family Process*, **22**, 85–98.

Bloom, B.L. (1985). A factor analysis of self-report measures of family functioning. *Family Process*, **24**, 225–39.

Byles, J., Byrne, C., Boyle, M., and Offord, D. (1988). Ontario child health study: Reliability and validity of the General Functioning Scale of the McMaster Family Assessment Device. *Family Process*, **27**, 97–104.

Epstein, N.B., Bishop, D.S., and Baldwin, L.M. (1982). McMaster model of family functioning: A view of the normal family. In *Normal family processes* (ed. F. Walsh). Guilford, New York.

Epstein, N.B., Baldwin, L.M., and Bishop, D.S. (1983). The McMaster Family Assessment Device. *Journal of Marital and Family Therapy*, **9**, 171–80.

Fuhr, R., Moos, R., and Dishotsky, N. (1981). The use of family assessment and feedback in ongoing family therapy. *American Journal of Family Therapy*, **9**, 24–36.

Green, R., Harris, R., Forte, J., and Robinson, M. (1991). Evaluating FACES III and the Circumplex Model: 2, 440 families. *Family Process*, **30**, 55–73.

Hampson, R.B., Beavers, W.R., and Hulgus, Y.F. (1988). Commentary: Comparing the Beavers and Circumplex Models of family functioning. *Family Process*, **27**, 85–92.

Jacob, T. and Tennebaum, D. (1988). *Family assessment: rationale, methods and future directions*. Plenum, New York.

Kinston, W. and Loader, P. (1984). Eliciting whole family interaction with a standardized clinical interview. *Journal of Family Therapy*, **6**, 347–63.

Kinston, W. and Loader, P. (1986). Preliminary psychometric evaluation of a standardized clinical family interview. *Journal of Family Therapy*, **8**, 351–69.

Lee, C. (1988a). Theories of family adaptability: Toward a synthesis of Olson's Circumplex and the Beavers Systems Models. *Family Process*, **27**, 73–85.

Lee, C. (1988b). Meta-commentary: On synthesis and fractionation in family theory and research. *Family Process*, **27**, 92–6.

Lewis, J., Beavers, W.R., Gossett, J.T., and Phillips, V.A. (1976). *No single thread: psychological health in family systems*. Brunner/Mazel, New York.

Marteau, T., Bloch, S., and Baum, J.D. (1987). Family life and diabetic control. *Journal of Child Psychology and Psychiatry*, **28**, 823–34.

Moos, R.H. (1990). Conceptual and empirical approaches to developing family–based assessment procedures resolving the case of the Family Environment Scale. *Family Process*, **29**, 199–208.

Moos, R. and Moos, B. (1976). A typology of family social environments. *Family Process*, **15**, 357–72.

Moos, R.H. and Moos, B.S. (1981). *Manual for the Family Environment Scale*. Consulting Psychologist Press, Palo Alto.

Moos, R., Bromet, E., Tsu, V., and Moos, B. (1979). Family characteristics and the outcome of treatment for alcoholism. *Journal of Studies on Alcohol*, **40**, 78–88.

Olson, P.H. (1985). Commentary: struggling with congruence across theoretical models and methods. *Family Process*, **24**, 203–7.

Olson, D. (1986). Circumplex Model, VII. Validation studies and FACES III. *Family Process*, **25**, 337–51.

Olson, D. (1991). Commentary: Three-dimensional Circumplex Model and revised scoring of FACES III. *Family Process*, **30**, 74–9.

Olson, D., Sprenkle, D., and Russell, C. (1979). Circumplex Model of marital and family systems: I. Cohesion and adaptability dimensions, family types, and clinical applications. *Family Process*, **18**, 3–28.

Olson, D., Russell, C., and Sprenkle, D. (1983). Circumplex Model of marital and family systems: VI. Theoretical update. *Family Process*, **22**, 69–83.

Reiss, D. (1981). *The family's construction of reality*. Harvard University Press, Cambridge, Mass.

Roosa, M.W. and Beals, J. (1990). Measurement issues in family assessment: the case of the Family Environment Scale. *Family Process*, **29**, 191–8.

Sigafoos, A., Reiss, D., Rich, J., and Douglas, E. (1985). Pragmatics in the measurement of family functioning: An interpretive framework for methodology. *Family Process*, **24**, 189–203.

Skinner, H.A., Steinhauer, P., and Santa-Barbara, J. (1983). The Family Assessment Measure. *Canadian Journal of Psychiatry*, **28**, 91–105.

Steinhauer, P., Santa-Barbara, J., and Skinner, H. (1984). The Process Model of family functioning. *Canadian Journal of Psychiatry*, **29**, 77–88.

Stierlin, H. (1972). *Separating parents and adolescents*. Quadrangle, New York.

Stierlin, H., Levi, L., and Savard, R. (1973). Centrifugal versus centripetal separation in adolescence: Two patterns and some of their implications. In *Annals of the American Society for Adolescent Psychiatry* (ed. S. Feinstein and P. Giovacchini). Basic Books, New York.

Westley, W.A. and Epstein, N.B. (1969). *The silent majority*. Jossey-Bass, San Francisco.

Part II

Clinical Aspects of the Family in Psychiatric Practice

3

Assessment of the family

The first two chapters of this book deal with some theoretical issues concerning family functioning. We now describe an approach to the assessment of families in clinical practice, providing a bridge between the previous chapters and those to follow in which family aspects of the major psychiatric syndromes are discussed.

The family assessment is an extension of two components of the conventional psychiatric assessment: the family history and an interview with one or more informants. An extended assessment can fit smoothly within the psychiatric interview and add a broader context to the resulting formulation. As with a conventional assessment, it may be built up over a number of interviews, the range and pacing of the inquiry varying with the particular circumstances of the case.

In this chapter we shall present a method of assessment which is consistent with the constraints of customary psychiatric practice, but which may, if the situation dictates, expand progressively in content and scope. We shall examine the circumstances in which a more detailed exploration of the family may be indicated, and how and with whom, this might be done.

Our starting point is the usual one for the general psychiatrist, an interview with the individual patient. We shall define four phases in the assessment:

(1) the history from the patient;

(2) a provisional formulation concerning the importance of the family in the case;

(3) an interview with informants (usually one or more family members);

(4) a revised formulation.

THE HISTORY FROM THE PATIENT

Undoubtedly, the most effective way to organize the family history is by constructing a family tree or genogram. This provides a striking, schematic representation of the structure of the family. By observing a consistent set of conventions, and with a little experience, a surprisingly large amount of information can be quickly grasped. Additional information can be added which summarizes important events and other qualities of the family. Scrutiny of the genogram also provides a fertile source of questions to be explored, and eventually of hypotheses concerning the relationship of the patient's illness to the family. Its assembly may be shared with the patient, thus helping to build rapport and fostering a shared interest in the context of the patient's problems. It may also, very successfully, suggest possibilities for change.

An excellent discussion of the genogram, its construction, interpretation, and uses in clinical practice, is presented by McGoldrick and Gerson (1985), to which the reader is referred for further details. Illustrations of the application of the genogram which draw on family trees of the famous (including the Brontes, Freuds, and Kennedys) enliven their book. We have adopted most of their conventions, and these are displayed in Fig. 3.1.

The following *personal details* might also be recorded for each individual: age, names, dates of birth and death, occupation, educational level, location, illness, functional information (poor functioning, problems, or successful functioning). Critical *family events*, for example, major transitions such as migration or other major moves, relationship changes, notable losses and achievements. Some of these will have been recorded under personal information, but it is often helpful to keep a separate chart of the family's chronology either in a margin or on a separate sheet. The *quality of relationships* can also be recorded. This information is at a more inferential level and care should be taken that adequate supportive evidence exists; be prepared to revise if new, contradictory information arises. Common representations are illustrated in Fig. 3.1.

Taking the family history

Who should be inquired about?
The answer to this question, we would suggest, flows from the nature and circumstances of the presenting problems. Useful principles guiding the taking of the family history are to work from the presenting problem to the broader context, from the present situation to its historical development, from 'facts' to inferences, and from non-threatening questions to more sensitive ones. An inquiry rarely omits an account of the current household, siblings and parents.

The presenting problem and the family
Commonly, questions are preceded by an introductory statement such as: 'in order to understand the problems better I need to know something

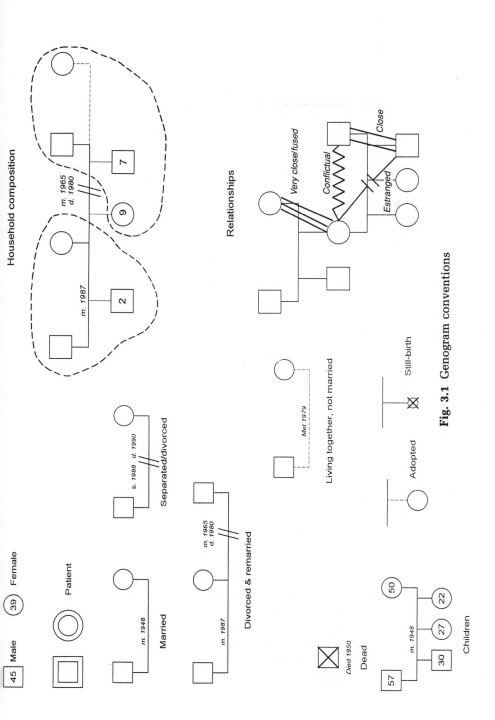

Fig. 3.1 Genogram conventions

of your background and your current situation. Who lives at home with you?'. However, this inquiry can be further enriched by questions which introduce interactional patterns and relationships: 'Who knows about the problem? How does each of them see it? Who thinks it is serious, and who thinks it is not? Who advised you to seek help? Has anyone in the family had similar problems? What happened to them? What treatment did they have? What has your family tried to do about the current problem? Who have you found most helpful, and least helpful? What do they think is the cause? What do they think should be done about it? What happens when you discuss it with your wife, or mother etc.'. The attitudes of key individuals in the family can be explored in this manner and shed light on the problem.

Relationship of the presenting problem to changes in the family

Useful questions aimed at understanding the current family context include: 'What has been happening recently in your family? Have there been any recent changes in the family (for example, births, deaths, illness, losses). Has your relationship with other family members changed since you have developed the problem? With whom, and how? Have relationships between family members changed?'. It is particularly important to ascertain whether the illness started around the time of an important family transition.

The wider family context

At this point a broader range of inquiry often flows logically; broader in terms of family members to be considered, as well as in the time span of the family's history. Parents' siblings and their families, grandparents, as well as a spouse's family may become relevant. Significant figures outside the family should not be forgotten. These may include care-givers and professionals.

Apart from information about the structure of the extended family, questions about how the family has dealt with transitions and significant events can be asked: for example, 'How did your family react when your grandmother died? Who took it the hardest? the easiest? How did migration affect your parents? Who adapted the quickest?'.

Relationships and roles can be explored at various stages during this inquiry, moving from relationships between the patient and other family members, to relationships between those other members. Especially close, distant, or conflicted relationships are of particular interest. An understanding of the 'roles' taken by family members may also be useful, for example, 'Who tends to take care of others? Who listens? Who is the spokesman for your family? Who needs to be taken care of? Who is rejected by the family? Who is successful?'.

Asking questions about relationships

Asking what a particular family member is like is important. However, we tend to gain a rather one-dimensional picture of the family solely from this type of question. Clinicians who work with families have evolved a style of questioning which aims to describe interactions and relationships. The clinician looks at differences between people, as described below, and responds to feedback from the family or individual, which thus shapes the next question, and so on in a circular manner, eventually leading to a joint, new understanding of the situation. This has been termed *circular* questioning; it derives from, and mirrors, the theory of circularity in systemic interaction, and the idea that it is differences between things that carry information. It is most effective when the questions are directed to exploring a particular hypothesis about the way the family works.

A *circular* question asks the patient to consider the perceptions, beliefs, or feelings of another person who has a relationship with the patient, and looks for differences between these other persons in their beliefs or feelings; for example: 'What worries your wife most about your problem? What worries your son most? What does your son say when your wife tells you that it is up to you to pull yourself out of it?'.

Several lines of questioning may reveal differences:

1. Pursuing sequences of interactions: 'What does your father do when you say the voices tell you to harm yourself? What does your mother say when he says that you should ignore them? What do you do when she says that to your father?'.

2. Detecting differences by 'ranking' responses: 'I know everyone is very worried that you may harm yourself. Who is the most worried? Who is the next worried? Who the least? Who is the most likely to try to do something when you talk about suicide?'.

3. Looking for changes in relationships before and after the illness: 'Since you have become ill, does your husband spend more or less time with you? Has he become closer or more distant from your daughter?'.

4. Hypothetical questions dealing with imagined situations: 'How do you think your relationship with your wife will change if you don't get better? Who in your family would find it most difficult if you were to tell them about your suicidal ideas? Who would be most likely to notice that you were getting better? Who would be the most sceptical?'

Examples will be seen above of *triadic* questions. These are useful in obtaining information about relationships and patterns of interaction which go beyond two people; for example: 'How do you see the relationship between your mother and yourself? How does your father see that relationship? Who would agree with you about how you see this relationship? Who talks more with your father, you or your sister? What would your mother say about what you have just told me if she were sitting here with us'.

Circular questions aim at addressing the dynamics of relationships. The general clinician may find them useful at some stages in obtaining the family history. However, a warning is in order here: both clinician and patient can easily become overwhelmed with information as a result of persistent questioning in this vein. Confusion rather than clarification may result. We recommend that this type of question be used principally when the clinician is exploring the effects on others of the patient's illness, and later, when he has specific ideas about the family's dynamics and wishes to check them for their consistency with new information.

MAKING A PROVISIONAL FORMULATION

Following the initial interview with the patient, two important questions about the family will commonly arise: (1) how does the family function, and (2) are there features of the family which might be related to the patient's illness?

How does this family work?

We propose a scheme for organizing one's thoughts about the family which builds from simple to complex observations. A family can be described at several levels which constitute a rough hierarchy:

(1) *structure*;

(2) *transitions and changes*;

(3) *relationships*;

(4) *patterns of interaction*;

(5) *the way in which the family works as a whole* (how it achieves continuity and balance).

(1) Structure
The genogram will reveal the structure (or shape) of the family. Some arrangements suggest that potentially difficult adjustments are required. These include, for example, single parent families, divorce and remarriage,

sibships with large discrepancies in age, or households which are unstable
in their composition. Unusual configurations invite conjecture about constraints and potentialities in such a family. Sometimes interesting repetitions
across generations may be observed; for example, single child families, a
preponderance of spinsters, premarital pregnancies, early deaths.

(2) Transitions and changes

Data will have been obtained concerning major changes and significant
events in the family. It is important here to consider the timing of
normative life-cycle transitions – births, departures from home, marriages,
and deaths. Have external events occurred coincidentally with important
predictable transitions, times at which the family may be more vulnerable?
The clinician will ask himself how the demands placed on the family by such
changes are being met.

(3) Relationships

Relationships refer to how family members interact with each other.
They usually involve pairs (or dyads), and their description is in terms
of closeness or distance, and their affective qualities – warm, tense, hostile,
and so on. Major conflicts may have been noted, as may estrangements or
unusually intense relationships.

(4) Patterns of interaction

Particular patterns of interaction may have become apparent. These go
beyond dyads. Triadic relationships are generally more informative about
how the family functions as a whole. In an important way, a third person
is integral to the definition of the relationship between another two.
For example, a third person may, in a variety of ways, play a role in
generating and managing tensions between two others. A conflict may
be detoured through a third person. This may prevent a resolution of
the thus submerged conflict. Appropriate generational boundaries might
be transgressed in such a case, for example, when a child acts in coalition
with one parent against the other.

(5) How the family works as a whole

At a higher level of abstraction, the clinician may form ideas about how the
family works as a totality. Higher organizing principles may have emerged.
Particular patterns of interaction, perhaps seen as a series of triads, may
be evident, which may have been repeated in more than one generation.
For example, mothers and eldest sons may have fused relationships, with
fathers being distant to both, while daughters and mothers-in-law are in
conflict.

  Idiosyncratic shared beliefs or traditions may be discerned which explain much of the way the family does things. These assume a superordinate position in the family's value system. 'Rules' governing the way members behave towards each other or to the outside world may flow from these. For example, a family may believe that 'you can only trust people in your own family; the outside world is hostile to us' and consequently may behave in a manner in which disagreements or conflicts are avoided at virtually any cost, and at the same time prohibit the seeking of support from an outsider. These beliefs and rules may be understandable given the family's past experiences (for instance, a Holocaust family in the example just given), but sometimes they appear to be 'habits' of mind or behaviour transmitted across the generations. When the family is faced with novel situations, these traditions and their associated rules may allow insufficient flexibility for arriving at a solution. The family may attempt to resist change in order to maintain balance within its customary modes of behaviour.

The way in which the family achieves equilibrium over time depends on many of the qualities discussed above. On the one hand there is a need for stability and continuity, on the other a need for change in response to normal transitions and external pressures.

At each level described above, evidence of difficulties in the family may be found and, if they are, the question will arise whether these play a role in the patient's problems.

Are family factors involved in the patient's problems?

Links between family functioning and the patient's problems take various forms. Some are immediately obvious, others may require a more extensive exploration of the context in which the illness has occurred. While recognizing this to be an oversimplification, we shall place these links in three categories. More than one will often, perhaps generally, be involved; however, we believe that a consideration of the role of each is helpful in providing an initial orientation to the context of the patient's illness. The three categories are:

(1) the family as reactive;

(2) The family as a resource;

(3) The family in symptom maintenance.

We do not in any way wish to imply that a family is to blame for a patient's illness, and we hope that a such a construction will not be placed by the reader on the following discussion, or indeed, in the volume as a whole. The family invariably will be involved to a greater or lesser extent in any member's illness. Our intention is to help the clinician to help the family in its distress, and to find ways in which its resources can be mobilized to assist the patient to recover, or at least to achieve the best outcome possible.

(1) The family as reactive

The patient's illness, or an exacerbation of it, may have occurred in a clear temporal relationship to a family upheaval which was independent of that illness. The upheaval may have unbalanced the family, disturbing its customary patterns of behaviour. These disturbances may still be evident as the family makes continued, perhaps so far unsuccessful, attempts to overcome its difficulties. The precipitant for the family's upheaval may be the illness itself. Fundamental rearrangements in roles or relationships might be required for stability to be maintained; these might have proved difficult to achieve. An escalating combination of both might be evident; the illness has occurred at a time of family stress, it stresses the family even more, and in turn this exacerbates the illness.

(2) The family as a resource

The family may be able to assist in the patient's treatment. This may be as straightforward as providing supervision of medication, encouraging attendance at an out-patient clinic, or detecting early signs of relapse, or it may involve the provision of a home environment structured in such a way as to help in the patient's convalescence or rehabilitation. The family may be able to call on friends or other agencies, professional or voluntary, to assist as well.

(3) The family in symptom maintenance

Interactions around the illness may inadvertently act to maintain it. There are two variants to be considered here. In the first, the patient's illness itself has become a way of apparently 'solving' a family problem. It thus could be said to assume an adaptive function, usually not a very effective one, but perhaps the best that can be achieved under the circumstances. An improvement in the patient would expose the family to challenges that it is presently unable, in fantasy or reality, to meet. For example, anorexia nervosa developing in a teenager who is due to leave home to attend a distant university may result in her needing to abandon this plan because she is unable to look after herself adequately. Were she to leave home, conflicts between her parents would become more exposed and her mother, with whom the patient is in an alliance against her father, would find herself apparently unsupported. The illness keeps the patient at home and involved in the parental relationship, while also providing a focus for a shared concern from others, giving the family a sense of unity.

In the second, maintenance of the illness does not solve a family problem although it may have done so in the past. A pattern of interactions has developed which persists out of apparent habit with a usefulness that has been outlived. An improvement in the patient would not expose the family to difficult challenges that it could not meet; instead it might lead to a ready readjustment and relief for all concerned. In the previous example, nine months later, the father's mother died. His wife subsequently offered him a feeling of closeness not experienced by him for many years, and their relationship improved substantially. Both parents, however, continued to treat their daughter as incapable of achieving greater independence, reinforcing her own uncertainties in this regard, and strengthening her fear of coping with life if she were recovered.

The categories of linkage outlined above are clearly themselves interlinked. The resources available for help in the family may be significantly determined by its burden of other problems, while an illness may come to have a stabilizing effect under some such circumstances. However, as an initial attempt to understand the family context of a psychiatric disorder, we suggest that a consideration along these three lines may be fruitful.

INTERVIEW WITH INFORMANTS

The clinician will now have made an initial assessment of the patient's problems and of the family context in which these have occurred. An interview with one or more informants is usually the next step in the psychiatric evaluation. The informant or informants are usually family members. We make no distinction between the informant interview and a family meeting. While the traditional interview has been with an individual seen alone, we suggest that under most circumstances more information will be acquired from a family interview, usually with the patient present.

There are a number of purposes served by the informant interview: to corroborate points in the history, to fill in important gaps, to determine influences operating on the disorder, and to recruit others to help. A family meeting is usually the most effective means of accomplishing these goals. Seeing an informant in relationship with the patient also helps to assess the 'objectivity' or reliability of the account given by that person of the patient's problem.

However, under some circumstances a family interview may not be appropriate.

Problems that might arise
The patient may refuse permission for family members to be interviewed. One should ascertain why, and weigh up the advantages and disadvantages

of proceeding in this way without attempting to change the patient's attitude. Reasons offered for not wishing the family to be seen include symptoms which have been kept secret, a belief that it is unfair to burden others with a personal problem, shame that one has needed to see a mental health professional, a fear that the family will be blamed, and sometimes suspicion or fear of family members. The reasons need to be discussed with the patient, particularly if the clinician has formed the opinion that the family context is important in the illness and that treatment will be significantly enhanced by their involvement. It is especially important to give an explanation for the clinician's request to see the family, usually in terms of the need to understand the problem more fully, that there may be much that the family can do to help, and that they may be grateful for information and advice. The absence of any intention to blame or criticize should be made explicit.

In most cases the patient will reverse his initial decision. In cases where there is a serious danger to the health or safety of the patient or others, and the involvement of the family is likely to be critical in preventing this, a patient's continued refusal may be overridden on ethical grounds. Otherwise, if the patient remains adamant, a refusal must be respected. The question can be raised again later when a more secure therapeutic relationship has been established.

Sometimes the clinician may decide to delay a request to see the family, for example, where the patient's engagement in treatment is tenuous and the clinician judges that a satisfactory working relationship might be jeopardized by bringing in others. Such a request might be perceived by the patient as signifying a lack of trust, confidence, or interest in his own perspective or experiences. Serious conflicts between the patient and family may heighten his sensitivities to these issues. The clinician may need to establish his credentials as a helper first.

Under some circumstances it may be appropriate to see informants without the patient being present. The risk of violence is often pertinent in such situations. It may be critical to establish some points in the history relating to the ascertainment of risk. An informant may not be able to speak frankly about this in the presence of the patient, particularly if there are fears for that person's or other family members' safety. Sometimes a psychotic patient may misconstrue remarks made by others which again may endanger his or their safety, or a patient with a chronic brain syndrome with failing intellectual powers may become distressed when his diminished capacities are discussed. It may be prudent to exclude the patient on such occasions.

When informants are seen without the patient, there is always the risk that the clinician will be told 'secrets' that cannot be ignored when the patient is seen again, but that, at the same time, cannot be discussed. Such situations prove awkward and raise important questions of the bounds of confidentiality versus the need to act in the patient's best interests.

Who should be seen?

The answer to this question depends on the purposes of the interview. We recommend that the standard meeting should involve all those living in the household who are thus likely to be mainly affected by the illness. However, departures from this norm are commonly necessary.

If the major aim of the interview is to fill in gaps in the history, a single individual is sometimes best placed to help. Young children are often excluded from an initial meeting unless they are closely involved with the illness. If the history indicates a significant involvement of non-immediate family members they should be invited. This also applies to persons outside the family. The more likely it seems that family factors are linked to the patient's illness, the more important it becomes for all significant members to be seen.

In considering whom to invite the patient's views should be sought. His advice might prove the most enlightening, and will provide an insight into whom he considers the key people in his life. A full explanation of the purposes of the interview should be given. He should also be asked what he thinks are the important issues to be covered, and may also be asked to predict what people will say and how they will behave.

Situations where it may be best to exclude the patient have been discussed above.

Finally, it must be recognized that whoever is seen or not seen, the clinician's entry upon the scene will have an impact on the family, and that if he is not careful, he will be drawn into roles or alliances dictated by family pressures rather than the needs for treatment. An understanding of how the family works may help to anticipate where the pitfalls may lie. Whether the patient or clinician is to invite the family can be negotiated between them. We generally favour the patient taking on this task, but sometimes it may be more appropriate for the clinician to do so. The clinician is likely to take the initiative when patients are disorganized in their behaviour as a result of their illness, or when they express a strong preference for him to do so. In the latter case, however, consider what this might imply about the patient's family relationships.

The family interview

It is still unusual for a family interview to be conducted as part of the normal psychiatric assessment. We therefore offer suggestions for those who may feel diffident in approaching this task. The interview affords an opportunity to gain more information about the illness, the patient's personality, and the family context. The clinician is able to meet and join with family members whose help he may later wish to enlist, and family interactions can be observed at first hand, particularly those surrounding the presenting problem.

Before the interview

By the time the family is seen, the clinician will have gathered much information from the patient and possibly others involved in the referral. A moment might be spent in thinking about biases that might have crept into the clinician's attitudes to the family as a result of the limited range of perspectives so far sampled. It is useful to try to adopt an open mind during the interview and to be aware of 'halo effects' introduced by previous discussions.

The clinician should also consider how the situation to date might dispose the family to attempt to draw him into alliances with some members. This is especially likely to happen where there are major conflicts or where desperately held views are under threat. Simply having seen the patient already, or having talked with others, may already have delineated possible coalitions. The clinician attempts to engage with the family as a neutral figure, whose only interest is to help solve the presenting problem.

The clinician will have a number of hypotheses about the family's relationship to the presenting problem. He might consider how these can be tested during the interview and have in mind particular questions and observations that must be included.

Starting the interview

In the initial 'social' phase of the interview introductions are made. Everyone's name and their preferred mode of address should be established. The clinician should be sensitive to the tone created by the family and adopt an appropriate style. The next step is an explanation by the clinician of the purpose of the meeting. Its details may be crucial to the success of his future dealings with the family. It is at this point that preconceived notions held by family members may be modified, potential barriers breached, and unhelpful anticipated alliances with the clinician placed out of bounds.

The clinician will usually give a brief account of the patient's presenting problem as he sees it. He will then say that in order to help the patient he needs to know more about the problem and that he is interested in hearing the views of everyone present about its nature and its effects on the patient and others. He may also say that he may need the assistance of those present to help the patient, and that they may also wish to ask questions and seek advice about how best to deal with the situation. It is crucial that a non-blaming stance is adopted. If a sensitivity to this is noted, an explicit statement is warranted: 'I want to stress that there is no evidence at all that families can cause an illness (or problem) like this. It is most inappropriate for anyone to feel guilty about this, although I know that despite what I say you may continue feel this way.'.

Those present will then be asked how they see the problem. There are differences in how experienced clinicians proceed here. Three practices are common. In the first, the order of inquiry follows the conventional hierarchy in families. The father in the family of origin of a young patient may be asked first, or the wife of a married patient. This approach demonstrates a respect for the hierarchy in the family which it may later prove important to reinforce. The second approach is not to direct the questions to anyone in particular and wait for someone to respond. This allows the clinician to see who is the family spokesperson or who feels most involved with the problem. The third approach addresses first, the family member of significance who appears most disengaged from the problem, yet whose involvement may be ultimately important. This signals to everyone that the person is seen by the clinician as having a significant role in providing information and perhaps later in helping. Each of these strategies is valid; we suggest that the clinician adopt the one which he believes is most appropriate, given what he already knows about the family, or the one with which he feels most comfortable. At any early stage everyone's views should be sought and eventually all should be given the feeling that they have been heard.

The presenting problem can then be explored in greater detail. As in the interview with the patient, this should be the initial focus with further questions flowing naturally outwards to broader issues. There will often be gaps in the history which need to be filled, and points requiring corroboration. It is a good idea to concentrate on facts first, before moving on to subjective responses.

Eventually, interactions related to the patient's symptoms will become a major focus – how the patient and others are affected by them, what measures have been taken to deal with them, who is most involved in various aspects of them, who is most distressed by them, who tries to comfort the patient most, how relationships in the family have changed since the illness started, how the illness has affected or been affected by family transitions, and so on. Circular questioning may be advantageous at this point in defining more clearly recurrent patterns of interaction.

The clinician may have ideas about how the problem relates to the way the family works. He can begin to test these out by asking questions and observing interactions. These ideas are usually kept private and not shared with the family. It is not helpful to put a hypothesis to the family and ask them whether it is true or not. It is better to ask about relevant details drawn from everyday episodes and behaviours, from which patterns may be inferred. For example, instead of asking about 'closeness' it may be preferable to ask about the amount of time spent together by particular members, whether intimate experiences are commonly shared by them, who helps with particular tasks, and so on.

Triadic relationships can be examined both by questioning – what does A do when B says this to C? – and by observation – what does A do when B and C reveal tensions during the session? The scope for circular questioning is greatly expanded when several family members are present. A third person may be asked to comment on what two others say to each other when a particular type of event occurs. This type of question often 'throws the family off their guard' by not asking them predictable questions to which they may by now have oft-repeated, stereotyped replies.

Information will be provided which elaborates the genogram. Observations may be made concerning the structure and functioning of the family – who makes what decisions, who attempts to control others' behaviour and in what areas, the quality of the dyadic relationships – closeness, conflicts, the main alliances – who supports who against whom, how clearly people communicate, how they attempt to solve problems, and their affective involvement with each other.

The discussion will gradually broaden to areas of family life not directly related to the patient's problem. Family beliefs and traditions will emerge, and sometimes, their historical origins. The rules of behaviour, how things are done in the family, will also become clearer.

Throughout the interview it is important that the clinician validates the comments and experiences of all of the family members. In addition to expressing his interest in the wider context of the patient's problem, he should also acknowledge the family's strengths and their efforts on the patient's behalf. Whether their attempts have proved effective in all respects or not, they have surely struggled to do the best they can under the circumstances.

The interview can be concluded with a brief résumé, in general terms, of what the clinician has learnt about the patient's illness and its family context. The family is thanked for coming and for demonstrating their concern for the patient. Usually the clinician will say that he now would like to think about what he has been told and that he may wish to see the family again. However, he may have already decided during the course of the interview that family involvement in the patient's treatment is desirable. He may then explain how they can help and arrange further meetings. The help may be fairly straightforward, as for example asking the family to provide supervision for the patient. An offer of on-going support and advice for the family may accompany this. On the other hand, the help requested may involve more complex matters relating to factors maintaining the illness, or at least interfering with its resolution.

REVISED FORMULATION

The provisional formulation can now be re-examined and modified as necessary. More information will probably be available at each of the levels of family description previously discussed. This is most likely to occur when the family interview has been well planned beforehand and its goals defined. This is particularly so for the elaboration of ideas about how the family functions as a whole.

We again suggest that the five levels of observation previously discussed be considered in turn (structure, transitions, relationships, patterns of interaction, and how the family functions as a whole) and that the question of how these might be related to the patient's illness be re-examined along the lines described earlier (the family as reactive, as a resource, or in symptom maintenance). By now the reactions of the family and the potential of the family as a resource will probably be fairly clear, and appropriate interventions can be planned.

The most difficult judgement will concern any role the family might have in maintaining the illness. We make some suggestions about what to look for. It is often useful to attempt to make a link between the illness and transitions in the family life-cycle. The following questions may help to discern such a link:

1. At the time of onset of the patient's symptoms was there a particular transition being experienced by the family?

2. Was this transition a particularly taxing one for this family, perhaps because of the way in which it had previously maintained balance? Did this transition impinge in a specially challenging way on family traditions or its habitual rules governing relationships? Were there coincident other transitions or unexpected events which might have made the family more vulnerable at that time? Are there structural aspects which might have made the transition especially difficult?

3. Has the patient's illness had the effect of preventing changes in relationships which would have been expected to occur as a result of a successful negotiation of the life-cycle transition? How does each family member derive relief through the avoidance of change?

4. Is there a special 'fit' between the patient's symptoms and the avoidance of change? Of the possible clinical expressions of the disorder, which is the most prominent and how might this have come about? Does the patient suffer from a disorder which is chronic but which would normally have been expected to resolve by now?

5. If there is no evidence that the illness is currently preventing change, is there evidence that at one time it may have have served such a role?

6. Is there evidence of specific family interactions around the illness that serve inadvertently to perpetuate it? Is there evidence that the family has not taken action that it might have been possible to take, or that it has been encouraged to take, to ameliorate the patient's illness? Is there a history of such behaviour, or was such behaviour observed during the family interview.

The clinician may find himself making affirmative responses to a number of these questions. Under these circumstances, a further exploration of the family context is indicated. By this stage, inquiries will become quite specific and will be aimed at trying to establish further support or a refutation of his hypotheses.

CULTURE AND THE FAMILY

This is an appropriate point to mention the important influence of culture on an understanding of the family. For our purposes a 'cognitive' definition of culture rather than one emphasizing behavioural practices is probably the most helpful (Keesing 1981):

A society's culture consists of whatever it is one has to know in order to operate in a manner acceptable to its members. Culture is not a material phenomenon; it does not consist of things, behaviours, or emotions. It is rather an organisation of those things. It is the form of things that people have in mind, their models for perceiving, relating, and otherwise interpreting them. Culture consists of standards for deciding what is, for deciding what one feels about it, for deciding what to do about it, and for deciding how to go about doing it.

This definition links culture with family belief systems, already described, and its entry into the assessment process commonly occurs at this juncture. The clinician will often find that he is struggling to achieve an understanding of the family in the face of two contrasting viewpoints – a 'relativistic' one, in which the family's structure, members' roles, belief systems, and so on, seem to be unique, and thus may be difficult to reconcile with his customary notions of family functioning, against a 'universalistic' one, in which similarities may be apparent because families share basic tasks such as procreation, child rearing, the differentiation of gender roles, and so on, and in which some common organizing principles should be evident. The advice of someone outside the family with an 'insider' perspective of the culture is often helpful.

Culture influences all aspects of family life – who constitutes the 'family' (for example, how many generations), which is the 'governing dyad' (husband-wife, father-son, brother-brother), parenting roles, the place of the sexes and of children in the family and in society at large, and so on. Culture is central in determining what is regarded as normal or acceptable, how symptoms are presented and to whom, and what kinds of treatments are appropriate or not. It also influences the family's interactions with outside organizations and institutions. For the clinician, such considerations may weigh heavily in decisions about whom to invite to a family meeting, to whom to address questions first, and of whom to ask particular types of questions. The clinician will also be alert to how he or she is construed by the family; gender, age, race, and position in the hierarchy of health professionals may be significant here.

Attention to the interface between the family culture and the host culture is important. Problems may relate not to the particular belief systems of the family *per se*, but to incompatibilities between these and those prevailing within the host culture and its institutions. Common examples involve adolescents who adopt the values of the host culture while the rest of the family remains embedded in its culture of origin, spouses who come from different cultures, and relationships with schools where conflicts about appropriate attitudes to child rearing may arise. It is also possible that problems are more likely to arise when traditional bonds and values are weakened than from their persistence. The interface between cultures may become even more complex if the clinician originates from a culture which differs from both the family's and the host culture.

Bear in mind also that problems may arise not because of cultural differences, but through stresses related to displacement, racism, poor social circumstances, unemployment, and poverty. Such factors are not simply reducible to the cultural background of the patient, and while the expression of a particular disorder may be shaped by his culture, an explanation for its appearance may not significantly entail cultural factors.

AN ILLUSTRATIVE CASE

We present a brief case history to illustrate our approach to the assessment of the family.

Mr Peter Dean, a 43-year-old married senior laboratory technician, was urgently referred by his general practitioner to the psychiatric out-patient department with a three month history of depressed mood, agitation, guilty ideas, and suicidal thoughts. His GP, who had known his family for many years, stated that Mr Dean had confessed that his life was 'a sham' and that he had been 'telling lies and stealing for many years'. He had shown his doctor a garage stacked with expensive items (tools, watches, cosmetics) either stolen or bought with stolen money. Short episodes of depression had occurred previously, but had not required treatment.

Mr Dean's father died when the patient was one-year-old. His mother remarried, to his father's brother soon thereafter, and they had two children. One of these, a son, died at the age of 30 following aspiration during an epileptic fit. Between the ages of 4 and 8, Mr Dean had been sent to a residential Catholic school because his mother was unable to look after him. There he had felt horribly lonely and deprived. When he was 8, he returned to his mother, now remarried. He was an average scholar and left school at 16. He then worked as a laboratory assistant, but through application and further part-time studies, he obtained a number of diplomas and for the previous 15 years had been a senior laboratory technician in a government scientific institute. His work was excellent and he was highly regarded.

Mr Dean married at the age of 22, his wife being a teacher at a special school for handicapped children. They had two children, a son, Peter, aged 21, and a daughter, Liz, aged 18. He described his family life as very happy. His mother and step-father lived nearby, and they had frequent contact. The GP called the family 'delightful'.

Mr Dean had a past history of epilepsy, with an onset at the age of 28. He was on anticonvulsant medication for 6 years, discontinued after a fit-free interval of 3 years. Both his stepfather and his half-brother had also suffered from epilepsy. His GP also reported that the patient had visited him fairly frequently over many years with a variety of somatic complaints, but that further investigations had never revealed any obvious causes for these.

His doctor described Mr Dean as a 'life-long worrier', but as also being 'extremely kind, considerate, and conscientious to the point of being a perfectionist'. He coped poorly when under stress, and ruminated excessively on negative comments made by others about him or his performance.

When asked about the stealing, Mr Dean stated that he had done this since childhood. He was remorseful, but also felt 'clever' when he got away with it. Stolen items were frequently given as gifts to family and friends. For as long as he could remember he had taken money from the till at his mother's drapery shop. Mr Dean also stated that he had never genuinely suffered from epilepsy; he had feigned the symptoms, as he had many others as well, in order to gain sympathy from others.

Mental state examination revealed a severely depressed man, with some psychomotor retardation. He was tearful and self-deprecating, claiming that he had wrecked his family. Concentration and energy were much impaired, and he complained of early morning waking associated with suicidal preoccupations. No delusions or abnormal perceptions were elicited, nor were there any cognitive abnormalities. He recognized that he was ill, but believed that he would never recover. His state seemed to him a just punishment for his past deeds.

In view of the severity of his depression and the risk of suicide, Mr Dean was admitted to hospital. In the course of further history taking and assessment, his family was seen. Interestingly, when asked who he thought should be seen first, he nominated his mother. Following an interview with her (and the patient), his wife and children were also seen (again in his presence). On the basis of the initial interview with the patient alone, and the two family interviews, the following family assessment was made:

(1) Structure

The genogram in Fig. 3.2 contains the salient points. His father, Peter, died at the age of 24 from tuberculosis after being decorated for bravery in the war. Peter's mother called him 'brave and handsome'. Note how the name Peter occurs in three generations. The mother's marriage to her former brother-in-law, and the death of Peter's half-brother with epilepsy have already been noted. The current household comprised Mr Dean and his wife, and Liz.

(2) Transitions and changes

The onset of Mr Dean's depression occurred around the time when Peter, his son, left home to live with his girlfriend. Liz was coming to the end of her school career and there was a good chance that she would soon be leaving home to pursue nursing.

(3) Relationships

Mr Dean had a troubled relationship with his mother. While understanding how difficult her life had been, he resented her placing him in an institution

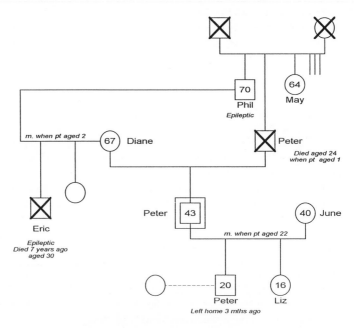

Fig. 3.2 The Dean family.

when he was a child. He had also felt cheated when she had remarried, and again later when she had directed so much of her energies into her business and looking after sick Eric. The way in which discussions about Mr Dean's father had always been avoided was quite amazing; in fact the patient did not know how old his father had been when he died, nor the cause of death. His mother had never shared her feelings about her dead husband with her son. When, during the interview together, he asked what he died of, Mr Dean immediately apologized to her, saying he did not mean to attack her. His aunt, May, had always told Mr Dean that his father was an exceptional figure and that 'you'll never be half the man he was'. He felt desperately humiliated by this. The dead father was an idealized figure who could never be talked about.

The patient's ambivalent feelings towards his half-brother, Eric, were openly expressed. He felt compassion for Eric's suffering, but also angry at him for being the subject of so much attention from his mother. He confessed to having stolen from Eric to deprive him of things he needed.

Mr Dean's relationship with his wife was a close one. They were both caring and loyal to each other, but Mrs Dean sometimes behaved more like a mother than a wife. The patient's relationships with his children were warm. He had showered his son with gifts and obviously wished him to have all the things he had never had. At one point his son remarked: 'I wish he would just be a father to me, instead of always trying to do things for me'.

Mr Dean's relationship with his GP was also noteworthy. He was very respectful to him, often asked him for advice, and in general terms, treated him like a father.

(4) Patterns of interaction

The way in which the nuclear family worked together to bolster up Mr Dean was most striking here. They sensed his vulnerability and worked together to detour any discussion which threatened to cast him in a negative light. This was evident both in the interview, and from the history, where following visits to his mother, he would often return home feeling irritable and require buoying up by his family.

(5) How the family works as a whole

The family shared the belief that a deprived background entitled the sufferer to special care and privileges. At the same time the losses that led to that entitlement were taboo subjects for discussion. It was one's duty to support the deprived person and at the same time to take gifts offered by them, even if they were unwanted and intruded on one's autonomy. The family, including his mother, immediately and unreservedly forgave Mr Dean for his stealing and deceit and urged him not to feel remorse. The death of Mr

Dean's father hovered like a ghost over the family (in concrete form in the name 'Peter'), but it could never be confronted. His mother's inability to grieve for him, perhaps contributed to by her marriage to his brother, had major repercussions for the next two generations.

We do not suggest that the depressive illness was caused by the family dynamics. However, we can postulate that its onset was related to young Peter's leaving home, and the likelihood of Liz leaving as well in the near future. For Mr Dean this represented the beginning of a disintegration of his family and the supportive matrix which sustained him. For his wife it meant that she had to shoulder the burden of his special privileges alone.

Mr Dean improved considerably with antidepressant medication and support in hospital. Following discharge there were four further family sessions during some of which the ghost of Mr Dean's father was discussed. Six months later he had fully recovered and was back at work. The stealing had apparently stopped.

REFERENCES

Keesing, R.M. (1981). *Cultural anthropology: a contemporary perspective*. Holt, Reinhardt and Winston, New York.

McGoldrick, M. and Gerson, R. (1985). *Genograms in family assessment*. W. W. Norton, New York.

Recommended reading

Burnham, J.B. (1986). *Family therapy*. Routledge, London. (This is a good introduction to ways of clinically assessing and thinking about families.)

4

Grief and the family

David Kissane

Grief is a universal emotional experience consequent upon a loss which has been portrayed throughout the centuries in great works of art, literature, and music. In the twentieth century the study of grief can be traced historically through the seminal contributions of Freud, Klein, Lindemann, and Bowlby. Arising out of their understanding of individual grief, an awareness of the significance of shared grief has developed in more recent times.

Mourning has a quite precise psychical task to perform: its function is to detach the survivor's memories and hopes from the dead.'

(Freud, S.E. 1912, p. 65).

Following the death of his father, his break with Adler and Jung, and the death of his brother Emmanuel in a rail accident, Freud completed his classic paper, '*Mourning and melancholia*' (1917), in which he contrasted grief with profound depression. In like manner, Melanie Klein's experience of personal loss was followed by creative achievement after her son died in a climbing accident and her daughter deserted her psychoanalytic group. Klein (1940) described the infant child's first feelings of concern for the loved mother, incorporating both a fear of loss and a longing to regain her: Klein most aptly described this set of feelings as 'pining'. She maintained that this capacity of the child to mourn is revived whenever grief is experienced later in life.

In the first systematic study of grief, Erich Lindemann (1944) observed 101 bereaved people who had either lost relatives in Boston's Coconut Grove nightclub fire or in World War II. He observed certain key features of acute grief, namely an initial sense of unreality, physical distress, preoccupation with images of the dead person, guilt and/or anger over the death, loss of customary patterns of behaviour and, on occasions, the development of symptoms similar to the deceased. He highlighted the phenomenon of anticipatory grief as well as describing both the delay and the distortion of grief that could occur when it became maladaptive. Summarizing the individual's experience of grief, Lindemann viewed the purpose of mourning as 'the emancipation from bondage to the deceased, readjustment to the environment in which the deceased is missing and the formation of new relationships.' This concise statement captured Freud's understanding and anticipated the development of Bowlby's attachment theory in the 1960s.

Aware of Lorenz's work on imprinting and Harlow's studies of the behaviour of infant monkeys, Bowlby (1961) described attachment behaviour as that which enabled a person to become or remain close to a preferred individual. Thus, a mature attachment provides a secure base from which a capacity to explore develops. Attachment is an integral part of human development, enduring throughout the life cycle. Linked with intense emotions of love and grief, it provides a clear survival value to the species. A person's primary attachment figure is usually a parent, most commonly the mother, but a spouse acquired later in adulthood replaces parents in the receipt of the strongest affectional bonds. However, it is clear that the family of origin constitutes the setting for the major constellation of bonds. Given the nature of such attachments in a person's life, the work of mourning following their loss is clearly represented as the undoing of these bonds. Thus attachment theory provides the most coherent theoretical model underpinning of the experience of grief.

Up to this point, the study of grief had focused on the individual person. In 1965, Paul and Grosser noted the dearth of work on shared grief. In the ensuing two decades, family therapists steadily made clinical observations on its significance, noting especially that when individual grief therapy failed, a systemic approach often succeeded. Altered family roles and interaction appeared to block individual grief resolution; their contribution was only understandable through a family approach. Lieberman (1978) summarized this clinical experience well in stating:

Clinically, it was common to discover that morbid grieving was a family pattern so that other members of the family needed to experience the forced mourning as much as the identified patient. When not included, family members tended to block the therapy by not allowing emotional expression of anger, grief and loss at home.'

(p. 163)

In 1991, Walsh and McGoldrick published the first volume on family grief, so ushering in a new era of the recognition of its importance. Since Lindemann's clear description of acute grief was available in 1944, we may ponder over why it took half a century for the family dimension to receive appropriate emphasis in this context? Study of the individual is substantially simpler than that of the family. A paradigm shift from that of the individual to systems thinking was also required in order to recognize the web of attachments particularly found within the family and to understand its impact on grief.

Loss takes many forms, its nature varying in the loss of home, job, health, money, relationships, and dreams as well as life. Even ostensibly desired change like retirement or marriage entails associated forms of loss. The loss of mental health too involves related grief for patients and their families, clearly exemplified by Alzheimer's Disease (Zarit and Zarit 1984) and schizophrenia (Miller *et al.* 1990). While we focus this chapter on death as our paradigm of loss, and then mostly involving the adult rather than the child, we acknowledge that this provides a model for other forms of family grief.

Terminology in this field is used variably and interchangeably. Bereavement describes the objective circumstance of a loss through death while grief refers to the range of emotional expression that ensues following loss. Psychoanalysts view mourning as a universal adaptational process that leads to satisfactory rapprochement with the lost person, with associated potential for a creative outcome. The lay public, however, view mourning as the socially sanctioned expression of grief following bereavement. Recognizing this diversity of the use of terms, we have chosen to restrict ourselves to 'grief'. The concept of anticipatory grief applies when a person begins to grieve in expectation of loss. This is also experienced by families when the death of one of their members is anticipated through terminal illness. Furthermore, as bereavement researchers have grappled with questions such as when grief should end, it has been accepted that the boundary surrounding the concept of normal grief is elusive. Similarly, a range of terms like pathological, maladaptive, and morbid are utilized to describe abnormal forms of grief. We believe that greater clinical utility is derived from the notion of adaptation than that of normality and hence commend the concept of maladaptive grief as most useful when problems are encountered with families. Let us focus now on what is known about family grief.

CURRENT KNOWLEDGE ABOUT FAMILY GRIEF

Understanding family grief derives from three disproportionate domains: clinical contributions (large, involving both case reports and family therapists' observations), systematic studies (smaller, with greater focus on child loss), and intervention research (sparse). A number of major themes emerge from this work which we now identify, starting with general patterns of family response to the death of a member.

Family response patterns

Adaptive family grief has been well described by Raphael (1984). The family functions with openness and an honest disclosure of feelings; positive and negative emotions are tolerated; members are intimate with each other and share their distress; and support is experienced through mutual consolation and care. These families operate flexibly (allowing for different rates of grieving), appropriately take on necessary tasks, and fill roles as needed. Growth occurs in intimacy and cohesion, with creative fulfilment arising from new interests and experiences. Such adaptive families do not need medical help beyond the respect involved in allowing them to proceed with their own grieving, and affirming its appropriateness.

In contrast, several *maladaptive patterns of family response* have been identified (see, for example, Lieberman and Black 1982; Bowlby-West 1983; and Raphael 1984). These are based upon various mechanisms including avoidance, distortion, inflexibility, and amplification of distress, all of which we briefly elaborate upon below.

(A) *Avoidant family response patterns* are common, vary in degree and include:

(1) silence about the deceased, regarding any discussion of the death as taboo;

(2) avoidance of intimate relationships with concealment of distress out of a fear of closeness;

(3) withdrawal from public engagement so that protective enmeshment enables the family to remain private;

(4) promotion of a family secret which upholds the family's pride at the expense of adaptively grieving; this may especially occur following a suicidal death;

(5) intellectualization of the experience – the family appears to 'do the right thing' but at the cost of sharing true feelings.

(B) *Distorted family response patterns* are commonly determined by a dominant member who leads the family to assume a stance that may prove adaptive for some but maladaptive for others. These patterns include:

(1) The family idealizes the deceased, promoting positive memories for some members but blocking others from resolving negative or ambivalent feelings.

(2) The family memorializes the deceased, keeping her alive and ever present, lest she be forgotten. The bedroom and possessions are preserved in every detail. The ghost of the deceased lives on, new relationships and change are impeded and chronic grief ensues.

(3) The family blames as a predominant response – this pattern is usually transmitted through the parents whose rigid control leads the family to

generate guilt and to seek out one or more persons on whom blame is directed.

(4) The family identifies with the deceased, perhaps adopting her social cause or perpetuating a sick role by taking on her symptoms. Modelling based on the deceased may be adaptive for some members but maladaptive for others, especially if imposed by a sense of duty to live up to the dead person's wishes.

(C) *Inflexible family response patterns*, occurring when the family has little capacity to cope with change include:

(1) The family insists life must continue as previously; since there is little flexibility of family roles, the deceased's role must be filled promptly for the system's maintenance. Such rigidity may place undue strain upon some members, inhibiting their grief and generating conflict.

(2) The family accentuates differences between members by rigid dependence on religious ritual and cultural tradition where different views and practices previously prevailed. A clash between family members may then occur, exemplified by religions that forbid a mourning ritual for a stillbirth.

(3) The development of inappropriate roles may flow for some family members from a lack of flexibility and adaptability on the part of others. Hence, children may be confined to restrictive relationships with a lonely parent or may be parentified following the loss of one parent.

(D) *An amplifying family response pattern* accentuates distress so that it reverberates through the family, disrupting cohesive bonds and prolonging grief. These response patterns include:

(1) The breakdown of a family for whom loss means chaos and disintegration; such a family has limited resources, their network of support also being deficient. They may have experienced separation, divorce, and mental illness in the past.

(2) Transgenerational rekindling of previously incomplete mourning of other losses of family members may serve to amplify current grief.

(3) Prolongation of grief through dependence on the ghost or commemoration of the person's death as a tragedy that must never be forgotten. This is complex with avoidance, distortion, and inflexibility all potentially operating to preserve a state of chronic grief in the family. On occasion, prolonged grief serves to avoid dealing with pre-existing

family difficulties; the family is thus bound together collusively through the extended grieving process.

The notion of identifiable patterns of family grief points particularly to transgenerational influences and the persistence of a family functional style.

Byng-Hall (1988, 1991) has approached the issue of family style ingeniously through his concept of the 'family script'. The script encodes the family's behaviour for future situations. Mechanisms like denial and identification with a replacement of a dead family member provide a means to replicate scripts which in turn generate problems for subsequent generations. 'Corrective' scripts may be adaptive, but if they prescribe behaviour in order to avoid painful experiences, they may also become maladaptive, inhibiting an effective family grieving process. Byng-Hall's notions are particularly helpful in determining long-standing family factors influencing the current response to loss.

Inherent in the notion of a family response to loss is the necessary recognition of individual differences in the rate of expression, both cognitively and behaviourally, of the grief. Such differences emerge particularly in parental grief, mothers generally experiencing more intense and longer grief reactions than fathers. This has been found in a number of studies covering Sudden Infant Death Syndrome, still-birth, and neo-natal deaths. The difference between the mother's and the father's grief may lead to misunderstanding and pave the way for marital disharmony. Sometimes a parent feels blamed for the child's death; at other times fathers accuse the mothers of prolonging their mourning. Few marital relationships appear unaffected by infant and childhood loss, but methodological difficulties (for example, lack of controls, small numbers) render it impossible to conclude that divorce is a more frequent outcome. Some couples become more distant while others grow more close in the wake of the loss (Dyregrov 1990).

In addition to these family response patterns, knowledge derived from both the clinical literature and systematic research on family grief are conveniently dealt with in terms of family structure and roles, communication and support, and the timing of the loss in the family life cycle.

Structure and roles

Various family sub-groups (for example, marital, parental, and sibling) have specific tasks and responsibilities in the context of bereavement; the loss of a family member also necessitates rearrangement of the bereaved person's role. Vollman *et al.* (1971) was among the first to observe how significant the role formerly occupied by the deceased was in this reorganization of the family's structure. She divided the role into instrumental, covering task-orientated functions such as being the bread-winner, or expressive, covering emotional functions like serving as the family's emotional 'barometer'. While the instrumental role might be difficult to fill, the expressive role was more vital for family equilibrium, its loss leading to disorganization and maladaptive behaviour. This latter role was also occasionally used to camouflage or resolve a conflict; its presence in these circumstances was crucial for the maintenance of family homeostasis. Other significant roles for the family include those exemplified by the parent of a young family, a 'special' child, or the family patriarch.

Communication and support

Communication in grieving families can be overt, including emotional expression, or symbolic, conveyed through consolation and support delivered by members to each other. Bowen (1976) emphasized the relevance of such communication as a key determinant of the ultimate pattern of grief. In his clinical experience, Bowen considered communication as either closed or open. When closed as a result of the patient's withdrawal into himself, the family members' avoidance of one another or the physician's technical jargon, adverse effects upon family grief ensued. In terms of the latter, Bowen strongly recommended that clinicians working with the terminally ill and the bereaved should be direct and explicit in their use of language. Although this recommendation has been repeatedly made by others in the field, the topic of communication patterns in the grieving family has not been adequately researched.

Closely linked with open channels of communication is the experience of support derived from both within and beyond the family. Vollman (1971) drew on her experience of working with families following a sudden, unexpected death. Cohesive families with an intact supportive social network adapted successfully whereas 'closed' families, often to the point of being socially isolated, were both at risk of poor outcome and resistant to therapeutic intervention.

Maddison and Raphael (1973) were also pioneers in concentrating on the relevance of support in grief in terms of the bereaved person's perception of the degree of helpfulness of her social environment. When family support was perceived as unhelpful and feelings were not expressed, maladaptive grief was more likely. Working with widows, it was noted that siblings and children generated more helpful exchanges than in-laws or mothers, with female-linked networks especially important. Sisters and their daughters provided considerable support but a widow turning to her own adult or adolescent daughter tended to derive most benefit.

Support, actual or perceived, has also been identified as an important factor influencing outcome after infant or child loss. In investigating the impact of perinatal death on families, Forrest *et al.* (1982) found that socially isolated women whose marriages lacked intimacy had a higher prevalence of psychiatric symptoms. Bereaved parents gained their most meaningful support from each other and from their own parents (Tudehope *et al.* 1986).

Reference to actual or perceived support raises the complex clinical and methodological question of what constitutes objective support. Is this possible to specify or are we bound to rely on the bereaved person's perception as the only meaningful aspect? It would seem to us that only a careful dissection of the person's experience, in conjunction with collateral judgements from family members, can shed light on this matter.

The experience of a lack of support brings us conveniently to the theme of loneliness. Several clinical researchers, including Parkes (1972) and Raphael (1984), have tackled loneliness as a salient aspect of bereavement. Glick and colleagues (1974) reported that 60 per cent of widows felt lonely at 12 months, while Lopata (1979) found that half the widows in his study described loneliness as their cardinal problem.

Large (1989) has contributed innovatingly to the topic in pointing out that families may exacerbate the loneliness of one of their members, particularly in a setting of unresolved grief. One person might bear an extra sense of emptiness whereas other family members were spared and thus able to function. The process however whereby some families move from a grieving state following an untimely death to chronic loneliness in one particular person has not been well studied.

The timing of loss within the family life cycle

The timing of a loss at different stages of the family life cycle generates particular sets of circumstances that are in turn associated with the family's management of grief and its potential complications. Carter and McGoldrick (1980) devised a well thought out scheme of the family life cycle as a framework for family therapy which has also proved useful in the study of grief. Hence, loss in the family can be discussed in terms of that experienced by unattached adults, the newly married couple, the family with young children, the family with adolescents, the family with adult-aged children, and the family in later life. In an excellent volume, the first to cover comprehensively the subject of family grief, Walsh and McGoldrick (1991) have emphasized the salience of the family life cycle in the wake of a loss through death. Within this broad context, death can be experienced as untimely, sudden, and unexpected, or anticipated and predictable.

Untimely death is exemplified by the Shanfield group's (1984) study of 24 parents' reactions to the death of an adult-child from cancer. Seventy per cent of the sample reported continuing grief two years after the death; such grief appears to be reasonably prolonged if it is experienced as untimely in terms of the family life cycle.

The effects of untimely death were investigated by Videka-Sherman (1982) through loss of a child. Unsatisfactory outcome, the development of marked depression, was related either to avoidance of grieving or to excessive preoccupation with the dead child. By contrast, satisfactory outcome for parents occurred when they adopted altruistic behaviour (associated with participation in a self-help group) or reinvested in another child or meaningful pursuit. Videka-Sherman postulated that replacement of the child was adaptive in resolving grief by parents actively directing their love and energy to another person. Two decades earlier, the question of the advisability of having a new child soon after a loss was substantially influenced by Cain and Cain (1964) in their caution against a 'replacement child'. Empirical research however does not support this view (Peppers and Knapp 1980; Murray and Callan 1988; Theut *et al.* 1989); a significant link has been found between a reduced level of depression and a successful pregnancy following loss. Theut *et al.* for instance, demonstrated a reduction in grief intensity in parents during the first six months of a new child's life. Although the parents were anxious about the viability of the infant, the concrete experience of this relationship was postulated to enable grief resolution of their prior loss to occur. The question remains as to what long-term effects might ensue for the replacement child himself.

The impact of sudden and unexpected loss upon a family was clearly demonstrated by Shanfield and Swain (1984) in a study of 40 parents two years after the death of an adult-child in a motor car accident. Although 90 per cent of the parents were still grieving intensely, with high levels of symptomatology (a third reported marked depression and loss of satisfaction with life), a substantial proportion indicated that the quality of family life had improved. This covered such aspects as closeness to their spouse and surviving children and the family's ability to resolve conflict and talk about emotional issues. Importantly, families rated as previously unstable because of the parents' ambivalent relationships with their children or the latters' experience of difficulties at the time of the accident, experienced more guilt and psychiatric symptoms than their previously stable counterparts. A high level of interdependence also predicted maladaptive grief. Poor family functioning thus appeared to be associated with the likelihood of maladaptive grief.

Other researchers have examined the effects of sudden, unexpected death by contrasting loss through Sudden Infant Death Syndrome with stillbirth and neonatal death. Retrospective data collection has marred much of this research and produced conflicting results, but in general the findings suggest that death from Sudden infant Death Syndrome results in more intense grief than when the death has been perinatal. It has been hypothesized that the longer the relationship between child and parents has existed, the greater the attachment, and therefore the more intense the grief in the event of sudden death.

Parrish *et al.* (1987) have further expanded our understanding of the problems associated with sudden death through a study of the care given to relatives by the staff of accident and emergency departments. A third of 66 survivors interviewed reported average or worse than average care. The need for explicit information was rated as an important means to achieve understanding of what had happened. Appreciating that the sudden, unexpected, and traumatic nature of deaths seen in casualty departments might predispose the survivors to maladaptive grief, Walters and Tupin (1991) developed useful clinical guidelines. Emergency department staff need to become adept in regularly informing relatives about the progress of the critically ill family member and supporting them during the notification of the death. Staff should facilitate emotional expression by the bereaved and offer support while they view the body. The family also usually require guidance about the need for autopsy, making contact with funeral directors, notifying other relatives, and a 'separation' process that helps them to sense when it is appropriate to depart. Provision of a means of follow-up is crucial.

In contrast to sudden and untimely death, adult loss, particularly that of the elderly, is commonly expected and synchronous with the family life cycle. The elderly develop deep attachments through marital interdepend-ence, become entrenched in certain roles, experience multiple loss, may be afflicted by profound loneliness, and have a personal awareness of their finitude. The family's tasks may be to assist their elderly relatives to adjust to new roles and possible changes in accommodation, as well as to promote the development of a variety of skills. In offering support, the adult children need to prevent the development of undue dependence upon themselves. As mourning in the elderly may continue indefinitely, the family needs to be particularly aware of the differential rates of grief among its members (Worden 1991).

Positive developments including personal growth may follow an expected adult loss. In a descriptive study in which Malinak *et al.* (1979) sought to identify adults' reactions to the death of a parent, the researchers noted the importance of the final farewell as a symbolic means of concluding the relationship with a terminally ill family member and preparing the bereaved for adaptive grief. Other investigators have followed caregivers prospectively through the illness and death of an elderly relative. Bass *et al.* (1991) for example reported that the perception of support received during this phase was more predictive of the pattern of family bereavement than support after the death. Thus, when family tension was evident during the care giving period, complicated bereavement was significantly predicted. When important sources of help to the caregiver were immediate kin only, the survivors later perceived greater difficulties in grieving. A health professional's participation in this group of important helpers protected against complicated bereavement. Bass and his colleagues suggested that the professional contribution might have benefited family members by providing instrumental assistance and promoting communication between them. Without this support, the children or siblings of the caregiver appeared to generate conflict over decision making, control, and perceived obligation to provide support. When clinicians become aware of family tension during a chronic or terminal illness, it is clear that preventive intervention with the family prior to death may prove valuable in enhancing subsequent adjustment.

An important and methodologically impressive project by Reiss (1990), the prospective study of the adaptation of both families and medical staff caring for end-stage renal disease patients, is a model of the research required on family grief in general. As the patient deteriorated and entered a terminal phase, Reiss observed the family's attempted realignment to reduce stress and their reinforcement of the patient-staff relationship. One dimension of the latter was the staff's increased authoritativeness in their support of both patient and family. Where the family failed to establish 'composure' in the terminal phase, bereavement became problematic. Acceptance of death played a pivotal role in reestablishing an equilibrium among the surviving family members. Furthermore, Reiss observed that family over-involvement in the chronic phase of illness, coupled with a rigid attitude towards compliance with treatment, was associated with family 'burnout', culminating in early death and vulnerability to complicated bereavement.

Finally, another family sub-group that may experience death out of synchrony with the family life cycle is made up of siblings. The grieving process of surviving children has been relatively understudied compared to that of their parents. The American psychoanalyst, George Pollock (1989), stands out as particularly observant of the sibling experience of loss. Among an intriguing collection of papers covering the work of three decades, Pollock portrays sibling loss in childhood as a family tragedy but asserts that the loss assumes a different meaning for each member; sibling loss, for instance, is more pathogenic for younger persons than for adults. In such circumstances, strained family relationships are reflected, *inter alia*, in parental overprotectiveness, formation of coalitions, parentification of children, blaming, competitiveness for attention, social isolation, and parental inability to emotionally support their surviving children. Sibling rivalry in this context is associated with guilt, anger, envy, and shame, all of which may distort the mourning process.

Pollock exemplifies these aspects by providing an intriguing insight into the adolescent Lenin whose older brother had been executed for his political idealism; he suggests that Lenin identified with, and sought to avenge his brother by opposing the political system, ultimately adopting the role of revolutionary. In a comparable childhood experience, Hitler lost four siblings, the last when aged eleven. Moreover, his bedroom overlooked the family cemetery. Pollock surmises that his sadism and arrogant claim to superiority arose through his family experience of sibling loss and the associated distorted mourning involved.

The above studies all point to the influence of the timing of loss within the family life cycle as a determinant of outcome. Whether the loss is expected or untimely is obviously crucial with the latter clearly increasing the risk of families developing maladaptive grief.

Let us finally turn to the third domain from which our current knowledge about family grief has been derived – family intervention studies.

Studies of family intervention

In their pioneering study on 'operational mourning', Paul and Grosser (1965) described a therapeutic approach in which families were encouraged to reflect on their loss, share associated feelings, and attempt to understand the impact of the death on themselves. They noted a high prevalence of inflexible family style, prominent denial of loss, and transgenerational influences. Although not reporting on the outcome of the study, they suggested that their approach was potentially efficacious. We hasten to clarify the term 'operational mourning' and its common synonyms 'forced mourning' and 'guided mourning'. All refer to a process in which family members share thoughts and feelings about their loss. This process may be likened to the exposure methods derived from behaviour therapy but it also involves facilitation of grieving through catharsis and clarification of family coping style.

Since the work of Paul and Grosser, only four studies have been done on the application of family therapy to grief. The findings are strikingly inconsistent, especially if we compare the work of Lieberman (1978) and Rosenthal (1980) on the one hand with that of Williams and Polak (1979) and Black and Urbanowicz (1987) on the other.

Lieberman (1978) reported on the treatment, using Paul and Grosser's approach, of 19 patients with maladaptive grief. Family participation facilitated the identified patients' acceptance of their loss. Twelve of the thirteen patients in which the family was involved benefited compared to three of the six who received individual therapy, a statistically significant difference.

Rosenthal (1980) examined the effectiveness of family therapy for maladaptive grief following the death of a child or adolescent in a sample of 15 families. Regression was a pathological feature found in a number of parents whereby they sought to have their dependency needs met by their children. Although the study was uncontrolled, positive results were obtained from 10 sessions of family involvement. Facilitating parents' greater tolerance of their own grief led to positive change for the whole family.

In contrast to this pair of investigations, two other studies on family intervention failed to show change. Williams and Polak (1979) applied a crisis intervention model to 32 families who had lost a relative in a motor car accident. A therapist accompanied the coroner's staff to meet the families within hours of the death and subsequently provided an average of five counselling sessions. The treated group was matched with two control groups, one bereaved and one non-bereaved. Apart from 'suddenness' of death, no assessment of risk factors for maladaptive grief was carried out. Rather than preventing morbidity, the investigators candidly conceded that they might have disrupted 'natural' mourning through their unduly early intervention; in fact, the contribution was perceived as intrusive by the families and this may have reduced their inherent potential to grieve effectively.

The second study, by Black and Urbanowicz (1984, 1987), involved 100 families. Six sessions were convened with the goal of promoting mourning for a parent who had died. No selection was made of at risk families. The attrition rate was high, with only 46 per cent of therapy families and 53 per cent of control families available for study at the end of two years. Outcome at a one year follow-up showed superior health and behaviour in children and less depression in surviving parents of the treated group. But these differences waned at a two year follow-up. However, more dropouts had behavioural problems after one year and these were assumed to have persisted. Hence, difficulties in engagement and compliance may have distorted the results. Other factors possibly reducing intergroup differences included the acquisition of a substitute parent through remarriage in some families and the 'therapeutic' influence of research interviews on the control group. The impact of the loss on the child was substantially affected by parental well-being and the capacity to support the child when he or she cried or talked about the dead parent. Reconstitution of the family through a second marriage may have hampered the child's emotional adjustment through avoidance of talking about the dead parent in the presence of the substitute parent. Although Black and Urbanowicz reported a null result, they were left with the conviction that intervention could be beneficial; on the other hand, they argued that replication studies were needed.

Apart from the intrusiveness of premature intervention as suggested by Williams and Polak (1979), studies have not reported deterioration resulting from family intervention. Gurman and Kniskern (1978) reported a 5–10 per cent deterioration rate for marital and family therapy generally but noted that such results were not likely to be published; this may be the case with the family therapy of grief. We are left, therefore, with inconsistent results from the family intervention work and corresponding uncertainty about which grieving families require professional help, and moreover, the optimal type of treatment.

MANAGEMENT OF FAMILY GRIEF

Having distilled out contemporary knowledge of the family dimension of grief, we are able to address those issues that are the everyday concern of the clinician confronted with the grieving family. Which families are at risk of maladaptive grief? How should the clinician proceed in the assessment of patients and their families? Which treatment approaches are applicable and efficacious? What preventive methods are available? In consolidating a rational approach built upon 'state of the art' knowledge, we address each of these questions in turn.

At risk families

Families at risk of maladaptive grief are those where factors about the death itself – its nature, timing, and context – combine with factors in the family's functioning – its cohesion, adaptability, roles, communication, emotional expression, and management of conflict – to generate strain upon the family's capacity to cope and respond effectively.

The nature of the death is problematic when its occurrence is sudden and unexpected or in some way shocking and traumatic. Hence, the suicide, homicide, violent death from any trauma, or the ambiguous experience of loss where a person goes missing, presumed dead, all involve horrifying fantasies about the experience of death that heighten the psychological distress involved. In contrast, while a lingering death may entail another form of distress, anticipation of loss prepares the family for the event with acceptance and understanding, so facilitating adaptive grief. Unnatural death commonly occurs in settings of conflict, anger, jealousy, or rage where relationships are based upon ambivalent or negative feelings and family life has often been chaotic and dysfunctional. These settings clearly produce an increased risk for maladaptive grief. Moreover, an unexpected death, although occurring in a relatively functional family, constitutes a major stress that may test its resources and at times exceed them. This is well exemplified by the 24 per cent marital separation rate in families where a child drowned compared with no marital break-down among matched controls experiencing a near-drowning accident, without a death (Nixon and Pearn 1977).

The timing of death is clearly relevant when one contrasts the expected loss of a grandparent, ill and frail for a period, with the death of a child or young adult. Untimely loss at whatever stage of the family life cycle imposes a strain upon the family's coping resources. Concurrent losses may also build up for families experiencing the deaths of several members within a brief time. Mourning of one loss, often incomplete, is rekindled and amplified by subsequent losses, heightening the intensity of distress and posing significant risk for dysfunction.

The context of the death in terms of the social and cultural world of the family merits attention since the influence of ethnic group, religious affiliation, and social class may all prevail. Thus, death within a migrant family often involves large geographical distances that prevent family togetherness as well as rekindling nostalgia for the mother country and ambivalence about departing from it years earlier. Death that occurs in war or in the context of torture is always untimely and unnatural. Death may also occur amidst community fear of contagion with corresponding ostracism, well illustrated by the AIDS epidemic.

Family functioning is a further key factor in determining whether a family is at risk of maladaptive grief. Where difficulties exist in a family's sense of cohesion, communication, emotional expression, adaptability over assuming roles and carrying out tasks, capacity to manage conflict and sharing interests, in short – to live well together – then the dysfunctional family style will engender increased risk for maladaptive grief among its members. Often the specific nature of the family dysfunction will prove problematic for grieving. Thus, where communication is poor, grief has less likelihood of being shared or where conflict is not readily tolerated, mutual blaming may ensue.

Systemic assessment of grief

The principles and procedures of family assessment in the case of grief are similar to those highlighted in Chapter 3. Assessment most commonly commences with the interview of an individual person with various concerns, problems, or issues. Although the clinician may utilize a number of approaches (for example, symptom list of the medical model, diagnostic classification of the psychiatric framework, developmental aspects based on psychoanalytic theory), a systemic perspective is an integral part of this inquiry.

When the clinician explores the issue of grief, posing a set of key questions constitutes this systemic approach. What changes did the last illness of the diseased bring about for the family? What was the deceased's role in the family? Who has filled the deceased's role? What was the nature of the family's loss? How has the family responded to the distress of grief? Who supports whom? Is the deceased talked about? How has the family coped generally? Whose needs are now greater and how have these changed over time? What happens in the event of a difference of opinion among family members? An additional, important inquiry focuses on the family's experience of, and dealing with, previous losses. The clinician strives to identify and clarify the specific grief issues for the family. It will often prove helpful if the clinician can map out the family's response pattern to the loss. Is the family, for example, avoidant, distorting, inflexible, or amplifying in their grieving style? The avoidant family will prefer not to discuss the deceased, conceal their distress, withdraw from the community, or rationalize the experience at the expense of sharing feelings. The distorting family will idealize the deceased, memorialize them excessively, blame others for the loss, or dutifully but reluctantly carry out the wishes of the deceased. The inflexible family will adhere rigidly to the family's previous routine, perhaps accentuating differences and conflicts between members and impose inappropriate roles upon its members. The amplifying family will open up old wounds, disintegrate through conflict or radical differences, or prolong grief through dependence upon the ghost of the deceased. Overlapping styles may be evident in a family at the same time although one issue will usually take precedence over others at any specific point.

Once a systemic framework for understanding the presenting problem has been accomplished, the clinician poses the question of how the family is to be recruited to an inaugural meeting. One strategy is to invite family members to attend in order to assist in the understanding and treatment of the presenting patient or 'to help in the situation'. Invariably, this collaborative approach is more successful in bringing the family together than a formal recommendation for family therapy. Clinical experience on the management of family grief (Gelcer 1983; Bloch 1991) suggests that the clinician should work with whatever family members are able or willing to participate. Therapy may then progress from the individual to the marital couple, through to the nuclear family, and, if necessary, to the extended family, each step facilitating the understanding of the relevant systemic issues. On occasion, transgenerational patterns will be clarified by the involvement of the extended family when they were not perceptible during the initial family assessment. Whatever the case, it is not the number of family members in the consulting room that is pivotal but rather the transmission by the clinician of the rationale of the systemic model to the family members present and through them to the rest of the system.

Treatment issues

The family intervention studies we have reported on do not offer much guidance for the treatment of the family in grief, particularly of the maladaptive kind. Williams and Polak's team (1979) caution about the intrusiveness of premature intervention should reassure the clinician that a therapeutic approach need not be made too promptly after the loss. Moreover, we can confidently rely on the adaptive potential of most families to complete the work of mourning with therapy only becoming necessary for those with symptomatic features or in the presence of high risk factors.

Paul and Grosser's (1965) model of 'operational mourning' is essentially one of open family communication, both through words and feelings, this facilitating the resolution of grief. It is the process, *par excellence*, for correcting the *avoidant family style* but this is not to suggest that we would advocate forced family mourning. Rather, it is often the acknowledgement of the barrier to grieving that the family has erected and the recognition of the sadness experienced by its members that will empower a family to proceed to grieve adaptively thereafter. Thus, instead of the family being forced to ventilate repeatedly their emotions in the consulting room, they are helped to observe and then adjust their process of grieving, the timing and style being left to their own decision.

Where a family has adopted a *distorted response pattern*, it is the search for other options and an appraisal of their respective advantages against the disadvantages of the current style that will most readily permit change to occur. Fortunately, there is often a sufficient range of resources within the family to pave the way for a new approach to loss. For instance, a family member may appreciate the emotional price to be paid in memorialization of the deceased and be able to comment both on its advantage and disadvantage to the rest of the family. Another may counter idealization by revealing negative attributes of the deceased, this initiative leading to a more integrated view. The cost of a dutiful and rigid pursuit of a family custom may be most readily manifested through the recognition of its deleterious effects across several generations (recalling the concept of the family script of Byng-Hall). And again, blaming may be highlighted transgenerationally and the anger often associated with it clarified in order that the latter is seen to be groundless or redirected more adaptively.

The *inflexible family response* arises as a protective style in families that fear change and sense that their capacity to cope with it is limited. Such families need explicit affirmation of their assets and help to acknowledge differences that prevail among their membership with associated potential for change. Mutual respect for diverse skills and the family's belief in themselves can be promoted to enable the tolerance of change as a positive step. Open discussion is encouraged about the roles filled by family members and their suitability so that corresponding tasks are linked to the appropriate generation with sub-system boundaries not crossed inadvertently.

Whenever *chronic grief* develops, avoidant, distorting, inflexible, or amplifying styles will have played a role in its promotion. Awareness of the systemic grief issues underpinning the system cluster in the presenting patient (for example, depression, somatization) is commonly the crucial insight that the clinician will share with the family; this in turn will constitute a challenge to explore their grief. The death may have occurred several years previously in which case it is only by sensitivity to the enduring impact of the loss and a thoroughness in tracing it through the family's life cycle that the clinician can aid the family in achieving a fresh understanding of its relevance.

The therapist may be reasonably confident that the systemic approach to treatment will enhance family cohesion and mutual support. Allman *et al.* (1992) found that supportive and engaging messages (for example, acknowledging sadness, complementing openness, reassuring that change would ensue, and confirming the value of working together) were frequently used in family therapy where grief issues predominated. The respect that the therapist manifests for the differential rate of grieving among family members will help to model an acceptance of such tolerance by all family members.

Individual and family therapeutic approaches may proceed in parallel, complementing each other and rendering each more potent. Thus, a bereaved family member who is severely depressed may require medication to achieve a 'biological hold' on his illness first before grief work can proceed at either individual or family levels. Thus, a systemic approach is woven into the therapist's spectrum of interventions and applied in an integrated and humane way.

Prevention

It is clear that most families, in the region of 80 per cent, grieve adaptively and will not require professional help. Their open and honest sharing of feelings and mutual support enable them to care for each other and adjust appropriately over time. Preventive interventions are not indicated for such families.

Families at risk, however, should be identified promptly and preventive strategies directed towards them. Those families in which the nature, timing, or context of the death render them vulnerable need recognition, as do families that have been customarily dysfunctional. Where a family demonstrates strain, tension, or conflict during the terminal illness of one of its members, intervention is usually appropriate. Similar principles of management apply to those utilized in the treatment of families grieving maladaptively. The goal is similar – to facilitate more adaptive family functioning so enabling them to realize their own potential to grieve.

CONCLUSION

Grief is an emotional experience that confronts all people, and, of necessity, their families. It offers an adaptational process whereby growth and creativity can be facilitated. However, failure to grieve adaptively is associated with significant morbidity, perhaps even mortality. An understanding of how the family grieves is a cornerstone to enabling adaptive grieving to be attained. Recognition of key family response patterns provides a systemic model on which approaches and treatment can be based. In our view, there is little doubt that those who understand and work through their grief can have their personal and family lives enriched; to put it succinctly – by appropriately mourning the dead, life itself can be enhanced.

REFERENCES

Allman, P., Bloch, S., and Sharpe, M. (1992). The end-of-session message in systemic family therapy: a descriptive study. *Journal of Family Therapy*, **14**, 69–85.

Bass, D.M., Bowman, K., and Noelker, L.S. (1991). The influence of caregiving and bereavement support on adjusting to an older relative's death. *Gerontologist*, **31**, 32–41.

Black, D. and Urbanowicz, M.S. (1984). Bereaved children – family intervention. In *Recent research in developmental psychopathology* (ed. J. Stevenson). Pergamon, Oxford.

Black, D. and Urbanowicz, M.A. (1987). Family intervention with bereaved children. *Journal of Child Psychology and Psychiatry*, **28**, 467–76.

Bloch, S. (1991). A systems approach to loss. *Australian and New Zealand Journal of Psychiatry*, **25**, 471–80.

Bowen, M. (1976). Family reaction to death. In *Family therapy: theory and practice* (ed. P.J. Guerin). Gardner, New York.

Bowlby, J. (1961). Processes of mourning. *International Journal of Psychoanalysis*, **17**, 317–40.

Bowlby-West, L. (1983). The impact of death on the family system. *Journal of Family Therapy*, **5**, 279–94.

Byng-Hall, J. (1988). Scripts and legends in families and family therapy. *Family Process*, **27**, 167–80.

Byng-Hall, J. (1991). Family scripts and loss. In *Living beyond loss: death in the family* (ed. F. Walsh and M. McGoldrick). Norton, New York.

Cain, A.C. and Cain, B.S. (1964). On replacing a child. *Journal of the American Academy of Child Psychiatry*, **3**, 443–56.

Carter, E.A. and McGoldrick, M. (1980). *The family life cycle: a framework for family therapy*. Gardner, New York.

Dyregrov, A. (1990). Parental reactions to the loss of an infant child: a review. *Scandinavian Journal of Psychology*, **31**, 266–80.

Freud, S. (1912). *Totem and taboo. Standard edition*, Vol. 13. Hogarth, London.

Freud, S. (1917). *Mourning and melancholia. Standard edition*, Vol. 14. Hogarth, London.

Forrest, G.C., Standish, E., and Baum, J.D. (1982). Support after perinatal death: a study of support and counselling after perinatal bereavement. *British Medical Journal*, **285**, 1475–9.

Gelcer, E. (1983). Mourning is a family affair. *Family Process*, **22**, 501–16.

Glick, I.O., Weiss, R.S., and Parkes, C.M. (1974). *The first year of bereavement*. Wiley, New York.

Gurman, A.S. and Kniskern, D.P. (1978). Deterioration in marital and family therapy: empirical, clinical and conceptual issues. *Family Process*, **17**, 3–20.

Klein, M. (1940). Mourning and its relation to manic-depressive states. *International Journal of Psychoanalysis*, **21**, 125–53.

Large, T. (1989). Some aspects of loneliness in families. *Family Process*, **28**, 25–35.

Lieberman, S. (1978). Nineteen cases of morbid grief. *British Journal of Psychiatry*, **132**, 159–63.

Lieberman, S. and Black, D. (1982). Loss, mourning and grief. In *Family therapy: complementary frameworks of theory and practice* (ed. A. Bentovim, G.G. Barnes, and A. Cooklin). Grune and Stratton, New York.

Lindemann, E. (1944). Symptomatology and management of acute grief. *American Journal of Psychiatry*, **101**, 141–8.

Lopata, H.Z. (1979). *Women as widows*. Elsevier, New York.

Maddison, D.C. and Raphael, B. (1973). Conjugal bereavement and the social network. In *Proceedings of Symposium on Bereavement*, November 1973. New York.

Malinak, D.P., Hoyt, M.F., and Patterson, V. (1979). Adults' reactions to the death of a parent: a preliminary study. *American Journal of Psychiatry*, **136**, 1152–6.

Miller, F., Dworkin, J., Ward, M., and Barone, D. (1990). A preliminary study of unresolved grief in families of seriously mentally ill patients. *Hospital and Community Psychiatry*, **41**, 1321–5.

Murray, J. and Callan, V.J. (1988). Predicting adjustment to perinatal death. *British Journal of Medical Psychology*, **61**, 237–44.

Nixon, J. and Pearn, J. (1977). Emotional sequelae of parents and sibs following the drowning or near-drowning of a child. *Australian and New Zealand Journal of Psychiatry*, **11**, 265–8.

Parkes, C.M. (1972). *Bereavement: studies of grief in adult life*. Tavistock, London.

Parrish, G.A., Holdren, K.S., and Skindzielewski, J.J. (1987). Emergency department experience with sudden death: a survey of survivors. *Annals of Emergency Medicine*, **16**, 792–6.

Paul, N.L. and Grosser, G.H. (1965). Operational mourning and its role in conjoint family therapy. *Community Mental Health Journal*, **1**, 339–45.

Peppers, L.G. and Knapp, R.J. (1980). Maternal reactions to involuntary fetal/infant death. *Psychiatry*, **43**, 155–9.

Pollock, G.H. (1989). *The mourning-liberation process*, Vol. 1. International University Press, New Haven, Connecticut.

Raphael, B. (1984). *The Anatomy of bereavement*. Hutchinson, London.

Reiss, D. (1990). Patient, family and staff responses to end-stage renal disease. *American Journal of Kidney Diseases*, **15**, 194–200.

Rosenthal, P.A. (1980). Short term family therapy and pathological grief resolution with children and adolescents. *Family Process*, **19**, 151–9.

Shanfield, S.B., Benjamin, G.A.H., and Swain, B.J. (1984). Parents' reactions to the death of an adult child from cancer. *American Journal of Psychiatry*, **141**, 1092–4.

Shanfield, S.B. and Swain, B.J. (1984). Death of adult children in traffic accidents. *Journal of Nervous and Mental Disease*, **172**, 533–8.

Theut, S.K., Pedersen, F.A., Zaslow, M.J., Cain, R.L., Rabinovich, B.A., and Morihisa, J.M. (1989). Prenatal loss and bereavement. *American Journal of Psychiatry*, **146**, 635–9.

Tudehope, D.E., Iredell, J., Rodgers, D., and Gunn, A. (1986). Neonatal death: grieving families. *Medical Journal of Australia*, **144**, 290–2.

Videka-Sherman, L. (1982). Coping with the death of a child. *American Journal of Orthopsychiatry*, **52**, 688–98.

Vollman, R.R., Ganzert, A., Picher, L., and Williams, W.V. (1971). The reactions of family systems to sudden and unexpected death. *Omega*, **2**, 101–6.

Walsh, F. and McGoldrick, M. (1991). Loss and the family: a systemic perspective. In *Living with loss: death in the family* (ed. F. Walsh and M. McGoldrick). Norton, New York.

Walters, D.T. and Tupin, J.P. (1991). Family grief in the emergency department. *Emergency Medicine Clinics of North American*, **9**, 189–206.

Williams, W.V. and Polak, P.R. (1979). Follow-up research in primary prevention: a model of adjustment in acute grief. *Journal of Clinical Psychology*, **35**, 35–45.

5

Affective disorders

Knowledge about the marital and family context of depression is in its infancy. Although a significant genetic contribution to bipolar disorder is now firmly established, comparatively little research has been done on the familial or genetic contribution to unipolar depression and dysthymia. Only a handful of studies have been carried out on the relationship between family environment and treatment outcome in affective disorders. A little more work has been done on the outcome of marital and family treatment; this is restricted mainly to couples therapy, reflecting the fact that affective disorders present primarily in adulthood. None the less, there is sufficient empirical and anecdotal material in the literature to allow useful guidelines for therapists about the types of marital and family interventions that are worthwhile. This is the main focus of what follows. Because they are substantially different conditions, depressive and bipolar disorders are discussed in separate sections.

DEPRESSIVE DISORDERS

Familial and genetic factors

A definitive study by Weissman *et al*. (1982) was based on 215 unipolar depressives and 1331 of their adult first-degree relatives. The results showed that depressive disorder occurred in the depressives' relatives more than twice as frequently than in those of the controls. This was confirmation of the consensus that had emerged from previous research, a consensus about which there is little disagreement.

The question has remained, however, of whether the excess of depressive disorder in depressed patients' relatives was attributable to familial or genetic factors. The most conclusive study in this context is that of Kendler *et al*. (1992), based on 1033 female twin pairs: the heritability of major depression was about 25–30 per cent, which represents a substantial but not overwhelming genetic contribution. This level is comparable to that found previously for coronary artery disease, stroke, and peptic ulcer, but substantially lower than that found for schizophrenia, bipolar disorder, and hypertension. Kendler *et al.'s* results suggest that environmental factors are more important than genes in influencing liability to depression. However, these influences do not appear to be familial; rather, they reflect personal issues such as stressful life events in adulthood. Kendler *et al.'s* findings reinforce an emerging consensus that parental rearing practices, social class of origin, and early parental loss do not generally have a major influence on the vulnerability to depressive disorder in adulthood.

Cultural factors

Although cross-cultural study of the family context of depression is rudimentary, this approach appears to be a promising means of elucidating the relationship between family function and depression. Apart from the heuristic nature of such investigations, they help to inform clinicians about those areas of family function that may require specific attention in different cultures. For example, Keitner *et al.* (1991) compared the functioning of families in Hungary and North America. In both countries, families with a member suffering from major depression reported poorer functioning than control families. However, the Hungarian families were most preoccupied with difficulties in setting or maintaining family rules and boundaries, whereas the American families were most concerned about problem-solving and communication difficulties.

The limitations of individual therapy

If individual therapies for depressive disorders were highly effective, there would be little need for marital or family dimensions. Several studies however show that conventional psychiatric treatment for major depression is generally of low effectiveness. Lee and Murray (1988), for example, followed-up, over 18 years, 89 consecutive admissions (62 women) to the Maudsley Hospital in 1965–66, all with a diagnosis of primary depressive illness. Mortality risk was doubled overall (including eight definite or probable suicides), and increased seven-fold for women under 40 at the index admission. Less than 20 per cent of the patients still surviving remained substantially free from depression, and over 25 per cent suffered severe, chronic distress or handicap. Ninety-five per cent had experienced at least one further episode of major depression, and nearly two-thirds had been readmitted to hospital at least once.

Kiloh *et al.* (1988) followed-up, for 15 years, 92 per cent of 145 depressed patients (106 women) admitted to a university hospital from 1966–70. Nine patients (8 women) had committed suicide, and only 20 per cent had remained well. Overall, the findings were very similar to those of Lee and Murray. Surprisingly, patients diagnosed as having 'neurotic' depression did little better than those with a diagnosis of 'endogenous' depression. The findings of Stephens and McHugh (1991) show the same pattern; they followed-up for a mean of 13.5 years 782 patients (54 per cent women) with a diagnosis of unipolar depression admitted to a psychiatric clinic from 1913–40. Twenty per cent substantially recovered; 35 per cent remained chronically depressed, spending almost one-third of the follow-up period in hospital. Overall, 56 per cent were readmitted at least once, and 14 per cent committed suicide.

There is, then, a surprising consensus about the long-term prognosis for those patients who are depressed enough to be admitted for in-patient psychiatric treatment. Only about 20 per cent substantially recover. One-third remain severely and chronically depressed, and many of the remainder continue to suffer indefinitely from troublesome symptoms.

Expressed emotion and recovery from depression

The value of family treatment in depressive disorder has been shown in studies using the Camberwell Family Interview to measure Expressed Emotion (EE) (Hooley and Teasdale 1989). EE reflects the extent to which close relatives express critical, hostile, or emotionally over-involved attitudes toward the patient (see Chapter 6). Several studies have now shown that higher levels of EE predict greater relapse rates after hospital admissions for depressive disorder. Hooley and Teasdale found that the best single predictor of relapse was the patient's response to the question 'How critical is your spouse of you?'. Further support for the predictive value of relatively simple measures of spouses' attitudes comes from a study by McLeod *et al.* (1992).

These findings imply that by asking the right questions clinicians can gain useful information about the role of the family in relation to a patient's depression. However, it is important to note that intervention studies examining the relationship between EE changes and outcome have not been done. Until such studies have yielded positive results, the role of EE in perpetuating depressive disorders will remain speculative.

The development of couples therapy for depression

It was the pioneering work of Weissman and Rounsaville (Rounsaville *et al.* 1979) that established the crucial relationship between marital adjustment and treatment outcome for depression. They studied a clinical population of 150 women receiving psychiatric treatment for depression. A majority of these women complained of marital problems. Thirty-eight of them were randomly allocated to a mean of 29 hours of individual therapy that focused on current interpersonal issues (rather than unconscious processes) and was aimed at helping the women to master their current social roles. Only 27 per cent of the women reported any benefits to their marriages after therapy. There was a marginal improvement in depression of women who had no marital disputes, or who resolved them during therapy. However, depression failed to improve or worsened in those where marital disputes persisted. Overall, the 38 women rated themselves more depressed after individual therapy than before, an unexpected finding substantiated by an increase of 50 per cent in the assessors' post-treatment ratings of depressive symptoms.

In light of these findings, Rounsaville *et al.* recognized that marital therapy might have a crucial role in depression, but concluded: '. . . it is frequently not feasible to treat more than one partner of a couple. For this reason, the present investigators have attempted to develop more precise techniques for dealing with the marital disputes of depressed patients' (p. 509). These attempts led to the development of Interpersonal Psychotherapy, an individually oriented approach which aims to examine and learn from the interpersonal context associated with onset of symptoms. Initial research on Interpersonal Psychotherapy suggested that it was of limited value in treating depression associated with marital conflict. This probably reflects difficulties depressed patients have in renegotiating their social roles without the active cooperation of their spouse. However, the large-scale NIMH Collaborative Research Program for the Treatment of Depression included Interpersonal Therapy as one of the three active treatments (Elkin *et al.* 1989). At the end of 16 weeks of treatment, there were few significant differences between the active and control treatments. Overall, Interpersonal Therapy was marginally superior to Cognitive Therapy but slightly less effective than imipramine plus clinical management.

The relationship between persisting depression and marital conflict has been well supported by work carried out after Weissman and Rounsaville's seminal contribution (see, for example, Beach *et al.* 1990). This created the impetus to develop marital therapies for depression, a number of groups having achieved considerable progress in this regard.

The work of Beach, O'Leary, and colleagues
The work of this group has been admirably documented in *Depression in marriage: A model for etiology and treatment* (Beach *et al.* 1990). We strongly recommended this volume which gives a particularly helpful overview of relevant theoretical work. Rather than repeat this here, we focus on aspects which are of importance to psychiatrists.

The Beach/O'Leary approach is explicitly a marital therapy and is restricted to those suffering from mild to moderate depression; the authors are clear about this. Patients with severe depression are unlikely to complete marital assignments, leading to '. . . an unhealthy focus on the depressed individual rather than the relationship . . .' (p.81). The authors recommend individual therapy, with an emphasis on somatic treatments, for the severely depressed. Only following substantial improvement is it appropriate for marital therapy to begin. Psychiatrists working with chronically and severely depressed patients may at this point lose interest in the marital approach of Beach and O'Leary, since it appears unhelpful in managing their most challenging patients. This rejection is premature because many chronically depressed patients do partially respond to somatic treatment, if only transitorily. Offering a marital approach at times of partial improvement may be crucial in achieving longer term benefit. Unfortunately, little data are available to confirm or disprove this.

The Beach/O'Leary approach is restricted to cases where *both* partners acknowledge marked marital conflict or dissatisfaction. In practice it is common for marital discord to be denied by one or both partners, even when evident to the clinician. Obviously, an offer of marital therapy here is likely to be rejected. Special techniques needed to engage such couples will be outlined below.

Beach *et al.* devote a chapter to a helpful overview of therapy. They mention obstacles to marital therapy such as persisting suicidal ideation, severe personality disorder, absence of gross marital discord (the issue of denied marital conflict is not addressed), limited marital commitment, and concealed agendas like extramarital affairs. Once a therapeutic alliance has been forged, therapy proceeds along behavioural lines, with the emphasis on goal-setting and homework assignment. Audiotapes are used to help couples learn communication skills, a central aspect of therapy.

O'Leary and Beach (1990) have compared their marital therapy approach ($n = 12$) with individual cognitive therapy ($n = 12$). The results show that marital therapy is only marginally better as a treatment for depression, but significantly superior for marital conflict. It is of interest that women with severe marital strife report increased marital dissatisfaction following individual cognitive therapy, a finding similar to that of Hafner *et al.* (1983). It should be noted that the subjects in O'Leary and Beach's study were obtained through newspaper advertisements, which creates some doubts about their comparability with clinical populations.

The work of N. Jacobson

Jacobson's approach broadly resembles that of Beach and O'Leary. It is explicitly behavioural marital therapy, focusing initially on analysis of behavioural exchange and then using training in communication and problem-solving to help couples resolve conflicts around issues like finance, sex, affection, intimacy, and parenting. One evaluation (Jacobson *et al.* 1991) compared behavioural marital therapy ($n = 19$), individual cognitive-behavioural ($n = 20$), and the two combined ($n = 21$) in married women who met DSM-III criteria for major depression. Fifty-eight per cent of subjects were solicited from newspaper advertisements but unlike those in Beach and O'Leary's research, they were selected on the basis of depression as a primary or exclusive problem. Many of Jacobson *et al.'s* subjects were not in fact maritally dissatisfied whereas this was a criterion for Beach and O'Leary.

This difference in clinical samples is crucial in explaining why marital therapy is effective for depression. In the Jacobson study, behavioural marital therapy was less effective than individual cognitive therapy as a treatment for depression, but more effective as a treatment for marital conflict. However, when analysis was restricted to patients who complained of the latter, behavioural marital therapy was clearly advantageous for both depression and marital conflict. This finding coincides with the work of O'Leary and Beach, strongly suggesting that a marital approach is useful for depression only in the context of marked marital dissatisfaction.

The Payne Whitney study

This well-known study by a team working at the Payne Whitney Clinic in New York compared Inpatient Family Intervention (IFI) plus standard hospital treatment with standard hospital treatment alone. Results for the 50 patients suffering from affective disorder (Clarkin *et al.* 1990) showed that only women with bipolar disorder benefited significantly from IFI, an effect that persisted to the 18 month follow-up. Men with unipolar depression surprisingly did better with standard hospital treatment alone. In both genders improvement correlated with a positive attitude of family members toward the patient.

Although the research team notes the striking nature of the sex difference in response to IFI, they are able only to speculate about the reasons for it. Depressed men, they suggest, may be so 'hypersensitive' to interpersonal stimuli that psychosocial interventions prove stressful. As well, the emphasis of male socialization on aspects like independence and self-sufficiency may predispose men to resist intervention that involves the family. Finally, men may be subject to more critical pressure from their families, especially with regard to the work role. Thus, there may be less flexibility and willingness to appraise family relationships.

Other work in the field

Another major study comparing marital with individual therapy specifically for depressed patients is that of Friedman (1975). Although methodologically flawed, it has been extensively cited. He compared four treatment conditions: amitriptyline plus marital therapy; amitriptyline plus brief individual supportive therapy; placebo plus marital therapy; and placebo plus brief individual supportive therapy. Both amitriptyline and marital therapy were significantly superior to brief individual therapy, but marital therapy was clearly superior to amitriptyline by the end of the 12-week course of treatment.

In addition to the above, numerous clinical reports and literature reviews are available which examine the role of marital and family therapy in depression. These are well summarized by Coyne (1988), Haas and Clarkin (1988), and Keitner and Miller (1990). Their main points can be summarized as follows:

1. It is difficult to establish the nature of any causal relationships between marital/family dysfunction and depression or *vice versa*. The idea of a direct linear relationship between them is probably simplistic. It is more realistic to consider a series of circular processes that gives rise to a mutually reinforcing negative interaction between patient vulnerability and family dysfunction.

2. Families of patients with major depression are commonly dysfunctional, their level of disturbance probably greater than that associated with bipolar disorder, schizophrenia, adjustment disorder, alcohol dependence, or physical disorders such as rheumatoid arthritis and cardiac disease. Over 40 per cent of adults living with a severely depressed person are themselves distressed to the point of meeting conventional criteria for therapeutic intervention.

3. Although family function usually improves when the patient's condition is ameliorated or remits, it often remains well below the level of non-clinical families. Problem-solving and communication are particularly deficient. Families tend to complain most about aspects of the patients' personality, particularly social withdrawal, irritability, negativity, and constant worrying. These enduring difficulties commonly give rise to more family concern than events surrounding acute episodes.

4. Relapse is associated with patients' perception of their spouses' attitude toward them. Those who perceive spouses as highly critical of them are more likely to relapse than those who view their spouses as supportive. The likelihood of relapse declines if family attitudes toward the patient become more positive. Persistence of negative family attitudes may increase the risk of suicide.

5. Although involvement of the family in the treatment of major depression is generally desirable, there are exceptions, especially when the patient is male. More work is needed to determine criteria for family/marital intervention as a supplement or alternative to individual treatment.

6. A range of family/marital treatments has evolved, but none have been rigorously appraised and several have had little or no independent evaluation at all. Promising approaches include: (a) behavioural marital therapy, as outlined above; (b) strategic therapy, a goal-oriented short-term marital approach that illustrates how the patient's symptoms may inadvertently be perpetuated by the couple's coping efforts, and helps them redirect their energies; (c) psycho-educational; (d) problem-centred systems therapy which involves as many of the depressed person's family as possible, and includes comprehensive assessment and modification of communication, emotional responsiveness, family roles, and discipline strategies; and (e) behavioural family therapy, involving detailed behavioural analysis, development of a therapeutic alliance with all family members, training in communication and problem-solving, and specific strategies for symptom management including medication.

How do psychiatrists currently treat depression?

There is a great deal of evidence for the desirability of a couples or family approach to the treatment of depression when it is associated with marital or family dysfunction. Several treatment models are available, and they can be used as a supplement or alternative to medication and other individual therapies. To what extent are psychiatrists using these models? A cross-national study by Glick *et al.* (1991) has helped to answer this important question. They studied a small number of cases in depth, so that their findings cannot be generalized with confidence. However, practices they identified in the USA appear representative. A striking finding was that, in spite of acceptance by treating doctors of the importance of the family during the acute phase of treatment, doctors themselves were very rarely involved in any family interventions, which were delegated to others. Overall, the quality of care was determined by how well the team was able to integrate biological and psychodynamic models. Although the research was done in leading academic centres, where quality of treatment is high, only about 50 per cent of what would be considered ideal care was delivered to patients, and only 37 per cent to families. Family interventions were largely ineffective: only 4 out of 24 families achieved successful resolution. There was little success in helping families to '. . . understand the relation of the symptoms to the family context, setting goals with or recognising major family problem areas (causing or resulting from the illness) or achieving psychoeducation . . . For the most part, at follow-up almost all of the families were either no better off, or worse, than at the time of the index episode' (p. 187).

Although this unsatisfactory situation relates to in-patient treatment for severe depression, it seems unlikely that clinical work in the out-patient setting involves the patient's family to any greater degree. The available evidence suggests that clinical management of depression remains largely individually centred.

Obstacles to delivering family therapy for depression

It is crucial to understand why family intervention is so uncommon in the treatment of depression, so that attempts can be made to improve the situation. We may note that much of what follows can be extrapolated to the treatment of other clinical states.

Few training programmes place much emphasis on marital or family therapy. The reasons for this are mainly historical: the emergence of Western psychiatry as an essentially medical discipline has meant that the individual doctor-patient relationship has acquired paramount importance. Although many training programmes attach great importance to psychotherapy, the emphasis tends to be on the individual approach. Marital therapy is rarely taught, and is considered to be of lower status than individual therapy. This view is reinforced by the observation that marital work is carried out by mental health professionals such as clinical psychologists, social workers, and psychiatric nurses. Organizations such as Relate (formerly the Marriage Guidance Council) in the UK are staffed mainly by lay people and although most are well trained and highly skilled, they are perceived as having low status.

Family therapy has achieved greater status than marital therapy but is seen as relevant to children and adolescents. This perception is reinforced by the established field of Child and Family Psychiatry, in which the terms family and child are virtually synonymous. Trainee psychiatrists are usually exposed to family therapy only during their rotation through a child/family psychiatry service. Moreover, most of what is written by family therapists concerns family function as it relates to children and adolescents.

The intrinsic limitations of marital and family therapy

As Beach and O'Leary explicitly state, marital therapy is practicable only in mild to moderate depression; it is useful when severe marital disturbance is acknowledged by both partners. Clinically, when this is denied by one or both spouses, it is difficult to ascertain whether or not this reflects reality. The depressed patient often forms the notion that any marital difficulty is caused by the illness, and is therefore reluctant to admit its presence. Moreover, depressed patients commonly feel vulnerable and dependent; in these circumstances, they fear alienating their spouse by acknowledging marital problems, since this may be construed as disloyalty. Spouses of depressed patients frequently regard themselves as exceptionally devoted, because they have remained in the marriage despite considerable complications surrounding the illness. An offer of marital therapy is therefore apt to be construed as an attack upon the marriage; this probably alienates the spouse further rather than increasing willingness to become constructively involved in treatment.

Invitations to family therapy in the treatment of depression in adults still living with their parents are often met with hostility. Parents may feel that the proposal implies they are to blame for their offspring's depression. The patient himself may hesitate, especially if he has come to view his depression as mainly biological in nature, existing independently of family or social factors.

There is little research on the extent to which couples and families reject a family approach. However, it is likely that only a minority of depressed patients with whom the possibility is discussed actually receive it. Even then, it is not uncommon for patient, spouse, or parents to insist that the illness remains the exclusive focus. Attempts to examine interpersonal issues may, unless made with tact and sophistication, result in drop-out. The more severe and chronic the disorder, the more likely are parents, spouse, or patient to cling to the idea that the depression is the pivotal target of treatment. This is a dilemma confronting clinicians who seek to manage severe depression in a marital or family context.

Practical approaches to the family treatment of depression

As we have seen, formidable obstacles confront the clinician who wishes to adopt a family approach. What follows is a distillation of the clinical literature to help as a guide. Because two-thirds of those having treatment for depression are women, we orientate this overview to them, and particularly to the wife as patient.

Engaging the couple

Suggesting to a depressed woman that her husband might become involved in therapy frequently evokes the following types of responses: 'He would get angry with me for suggesting it'; 'He thinks therapists are fools who know less about living with a depressed person than he does'; 'I don't want to involve him because it's really my problem'; and 'I won't even ask him because I know he'll refuse to come'.

A wide range of hidden agendas underlie these responses, some of which have been outlined previously (Hafner 1986). The main themes in depressed women relate to: fears of abandonment by the husband; the wish to keep a private, exclusive relationship with the therapist; and dynamic issues relating to past attempts at individual treatment that engendered hostility in the husband toward the therapist or the idea of therapy. Often, sensible discussion and reassurance allow an invitation to be extended to the husband, and a courteous letter usually works (although we are aware of cases in which husbands tear up the letter without opening it!). It is vital to understand that most patients who refuse an invitation to marital therapy after rational discussion have good reasons for doing so. Even though these may not be divulged, they must be respected. A depressed patient should not be pressured or cajoled.

Marital or spouse-aided therapy?

A crucial issue is what to call the therapy. Using the term marital therapy will often alienate couples, for reasons outlined earlier. However, research employs this term. A way of avoiding the dilemma is to use the term 'spouse-aided therapy'. This originated during the evolution of a couples approach which is explicitly *not* a marital therapy, and which proved significantly superior to individual therapy (Hafner *et al.* 1983). In essence, the spouse is invited to take part as collaborator, co-agent of change, co-therapist, or therapist's assistant (the terms are fairly synonymous). Few spouses can resist the opportunity. Critics point out that this approach is likely to confirm the patient's dependent, inferior position and may aggravate marital factors reinforcing the depression. The risk is worth taking if it is the sole way of initiating couples therapy. Such developments can be pre-empted; a particularly helpful strategy is to establish treatment goals which require active problem-solving on the part of both husband and wife. There is clinical consensus among those who work in this difficult area about the salience of setting clearly defined goals that require active collaboration of both partners (Coyne 1988).

Suicide, personality disorder, and marital conflict

The risk of suicide is markedly elevated in severe, persisting depression. Treating suicidal patients in a couples setting is a particular challenge, because of the dynamics involved. However, suicide is more likely to occur if the depressed patient sees her spouse or family as critical and unsupportive (Keitner *et al.* 1990). This underlines the importance of improving family ties. Patients with combined personality disorder and depression are at risk for suicide, while the presence of personality disorder markedly increases the likelihood of marital disturbances. At least a third of patients with major depression are personality disordered in one or another way (Haas and Clarkin 1988). Thus, it is common to find a close link between threat of suicide, personality disorder, and marital conflict.

Probably the most critical point in managing suicidal patients in a marital or family setting is to allow them to talk openly about their suicidal thoughts, including any plans, and especially those involving hoarded medication, or firearms. There are numerous practical and dynamic reasons for suicidal patients to resist such openness but if the therapist creates a suitable framework the opportunity then emerges for suicide to be placed firmly on the clinical agenda.

Medication and family treatment

Clinicians who use family therapy for severe, persisting depression generally regard antidepressant medication as vital. Combining medication and family therapy often enhances both. Once patient and family appreciate the rationale of medication and its likely side-effects and benefits, compliance is enhanced. Response to medication also improves the patient's motivation to set treatment goals and to work at them. Increased energy, concentration, and interest facilitate working towards achievement of these goals. However, even severe, persisting depression can improve substantially without medication if family or marital therapy succeeds in resolving key elements of conflict.

Contraindications to marital and family therapy

It is clear from the discussion so far that work in marital and family therapy for depression is at an early stage. Although the presence of marital conflict is an indication for a family approach, we know little about contraindications. The work of Glick *et al.* (1991), as previously highlighted, suggests that men are less likely to respond than women. The prognosis of severe depression when combined with marital conflict is generally poor, and if persistent, the likelihood of suicide is in excess of 10 per cent. In these circumstances, clinically informed attempts at family intervention are unlikely to be detrimental. An attempt however to predict the outcome of spouse-aided therapy for severe, persisting psychiatric disorder, including depression, suggests that high levels of outwardly directed hostility in *both* partners is associated with a poor outcome. Outcome is also poor when both partners rely heavily on a biological model to explain the patient's disorder. Such a viewpoint is not unexpectedly associated with denial of personal and interpersonal aspects, especially by the spouse.

BIPOLAR DISORDERS

In a prospective study comparing 73 bipolar with 66 unipolar depression patients, Harrow *et al.* (1990) found that the bipolar patients had a worse prognosis. Their findings were similar to those of other studies, and they concluded that '. . . many hospitalized manic patients have a severe, recurrent and pernicious disorder'. They suggested that lithium carbonate is a less effective treatment for this group than generally believed. Thus, the challenge of treating severe bipolar disorder seems as great, or greater, than that of managing severe depression.

Patients with bipolar disorder are probably less likely than their depressed counterparts to receive marital or family therapy, despite evidence suggesting its importance in overall treatment. However, data are mainly uncontrolled or anecdotal, and only in the work of Clarkin *et al.* (1990) have patients been randomly allocated to either family or individual treatment. As previously noted, women with bipolar disorder responded better to family intervention, which was psychoeducational in nature. Additional support for family intervention comes from a study by Miklowitz *et al.* (1988) of 23 patients while manic. Measuring EE in 36 key relatives, it was found that negative attitudes toward the patient were highly predictive of relapse, this independent of medication compliance, treatment regime, baseline symptoms, demographic data, and illness history.

Family dynamics in bipolar disorder

Knowledge about marital and family aspects of bipolar disorder revolves around psychodynamic issues, about which there is little consensus. The disparity in findings may be due to sample differences, since most psychodynamically orientated studies are based on small numbers. More-over, since the divorce rate among bipolar patients exceeds 60 per cent those who remain married and hence lend themselves to study, are likely to exhibit unusual features. This may explain frequent references to the 'symbiotic' nature of the marriages, the partners having a high degree of mutual influence.

A second consistent theme is the suppression of anger by both partners. In the spouse, this revolves around events that occur when the patient is manic. Given the marked social and financial problems related to mania, it is not surprising that spouses harbour resentment and anger. However, because the behaviour is largely attributable to an illness over which the patient has little control, the spouse feels guilty about expressing anger to him. The latter in turn feels guilty about the consequences of his manic behaviour; and, especially when depressed, feels dependent on his spouse, with fears of being abandoned by her. For these reasons the patient inhibits direct expression of anger. This mutual suppression of anger makes open, honest marital communication difficult, and impedes constructive problem-solving. Moreover, anger tends to be expressed indirectly, contributing to a steady elevation in marital and intrapsychic tension which in turn increases the likelihood of further manic episodes.

Family management of bipolar disorder

A central theme in management is the balance between biological and social factors. Although psychosocial factors probably contribute only about 25–30 per cent to prognostic variance, this contribution is vital in that it represents a major area for intervention. Moreover, appropriate strategies probably enhance medication compliance, thereby sharing the variance attributable to biological factors. Family interventions have been little used for two reasons. First, the well-established genetic contribution has led bipolar disorder to be considered a biologically determined illness for which psychosocial intervention is irrelevant; the role of lithium has reinforced this viewpoint. Second, there is little professional awareness of family intervention designed specifically for bipolar disorder. Professional views are commonly internalized by patients who then see the condition as a medical illness over which they have little or no control. This perspective is reinforced by attendance at specialized lithium clinics and regular monitoring of drug levels.

Davenport and Ackland (1988) have evolved family techniques that consider these issues. Their approach is relevant because it is often initiated during an acute phase; family interviews may be conducted with patients disturbed enough to be in seclusion. Not uncommonly the patient's behaviour improves dramatically in the presence of family members. Most families are keen to be involved, and, once established in a working relationship with the professional, a useful alliance may endure for years.

Family sessions are pragmatic, although informed by psychodynamic, strategic, and psycho-educational principles. During initial meetings the therapist gains information that facilitates management, and family members learn about the disorder systematically. Interventions are designed to change maladaptive defences, improve family communication, and introduce more effective coping. Ventilation of anger and frustration is in itself an element that helps to ameliorate guilt.

Davenport and Ackland emphasize that medication compliance is substantially enhanced by the family approach, a finding confirmed elsewhere, although no empirical evidence for this effect is available. However, they do not mention targeted medication, or training families to detect warning signs of depression or hypomania. Admission to hospital is more frequent in those who are poor at recognizing these signs (Joyce 1985). Considerable opportunities exist for teaching patients and families to work together at detecting them. This allows additional medication to be introduced at optimal times rather than administered continuously, with the likelihood of increased compliance. Inviting patients and families to assume more responsibility may consume more professional time initially, but if it reduces the time required in the medium- to long-term, then it is well worth-while.

Psychiatrists trained in family therapy will probably find the approach of Stierlin *et al.* (1986) and Weber *et al.* (1988) valuable. It is the product of a group of experienced clinicians who have collaborated in this area for several years. Thus, although support for their approach is anecdotal, the ideas and methods warrant serious attention. Although both papers are easy to read and refreshingly free of jargon, the ideas they contain are difficult to summarize briefly; we strongly recommend them.

POSTPARTUM DEPRESSION

A comprehensive review of postpartum depression in *Motherhood and mental illness* (Kumar and Brockington 1988) includes a summary of research on the marital relationship. Of five prospective studies four found postpartum depression more likely to occur in women who had reported a poor marital relationship during pregnancy. Boyce *et al.* (1991) measured interpersonal risk factors in 149 non-depressed women during pregnancy; dissatisfaction with the spouse was the most robust predictor of maternal depression one and three months after childbirth. At six-months postpartum, the level of depression was best predicted by interpersonal sensitivity and the degree of depression at three months. This well-designed investigation suggests that a supportive spouse is important in the first three months after childbirth. Once postpartum depression develops, it seems to persist in women with vulnerable personality traits.

Appropriate counselling during pregnancy diminishes the likelihood of postpartum depression (Kumar and Brockington 1988). Although involving husbands in prenatal counselling adds to its effectiveness, this requires empirical confirmation. We know even less about the husband's role after depression has developed. Once again, it is likely that there are benefits of helping the husband to be more supportive during this critical phase of the family life cycle. But central issues remain unclear. For example, should husbands be recruited to share as much as possible in looking after the new-born, or should this be discouraged on the basis that it undermines the mother's attempts at developing her own nurturing role?

SEX ROLE ISSUES IN DEPRESSION

Sex role issues are probably salient in any consideration of depression (Hafner 1986). Clinically, they manifest in two ways. First, married women with severe depression often have a traditional and unusually fixed view of sex roles. By virtue of assortative mating their husbands harbour similar attitudes. The women may be excluded from activities that protect against depression, particularly paid employment outside the home. Second, women with traditional but less rigidly held attitudes may experience high levels of conflict over which sex-role behaviours are most appropriate. The conflict concerns dependence, vicarious achievement, and sacrifice of personal goals versus independence, direct achievement, and active pursuit of personal objectives. Conflict over sex roles may contribute to depression which in turn consolidates dependency. Although dependency reduces the conflict temporarily, it also negatively reinforces depression and helps to perpetuate it.

Attention to sex role matters during marital and family therapy allows changes in patients' attitudes that may ameliorate the depression. Obviously, therapists must take care not to impose their own views about sex roles (see Chapter 14). Helpful interventions include provision of information about the effects of sex role stereotyping. This introduces an element of choice, allowing patients and their families to make appropriately informed decisions. Once the decisions have been arrived at, the therapist often has a role, albeit challenging, in aiding the family to implement changes.

REFERENCES

Beach, S.R., Sandeen, E.E., and O'Leary, K.D. (1990). *Depression in marriage: a model for etiology and treatment.* Guilford, New York.

Boyce, P., Hickie, I., and Parker, G. (1991). Parents, partners or personality? Risk factors for post-natal depression. *Journal of Affective Disorders*, **21**, 245–55.

Clarkin, J.F., Glick, I.D., Haas, G.L. *et al.* (1990). A randomised clinical trial of inpatient family intervention. Results for affective disorders. *Journal of Affective Disorders*, **18**, 17–28.

Coyne, J.C. (1988). Strategic therapy. In *Affective disorders and the family* (ed. J.F. Clarkin, G.L. Haas, and I.D. Glick). Guilford, New York.

Davenport, Y.B. and Ackland, M.L. (1988). Management of manic episodes. In *Affective disorders and the family* (ed. J.F. Clarkin, G.L. Haas, and I.D. Glick). Guilford, New York.

Elkin, I., Shea, M.T., Watkins, J.T. *et al.* (1989). National Institute of Mental Health treatment of depression collaborative research programme. *Archives of General Psychiatry*, **46**, 971–82.

Friedman, A.S. (1975). Interaction of drug therapy with marital therapy in depressive patients. *Archives of General Psychiatry*, **32**, 619–37.

Glick, I.D., Burti, L., Minakawa, K. *et al.* (1991). Effectiveness in psychiatric care II. Outcome for the family after hospital treatment for major affective disorder. *Annals of Clinical Psychiatry*, **3**, 187–98.

Haas, G.L. and Clarkin, J.F. (1988). Affective disorders and the family context. In *Affective disorders and the family* (ed. J.F. Clarkin, G.L. Haas, and I.D. Glick). Guilford, New York.

Hafner, R.J. (1986). *Marriage and mental illness: a sex roles perspective.* Guilford, New York.

Hafner, R.J., Badenoch, A., Fisher, J. *et al.* (1983). Spouse-aided versus individual therapy in persisting psychiatric disorders: A systematic comparison. *Family Process*, **22**, 385–99.

Harrow, M., Goldberg, J.F., Grossman, L.S. *et al.* (1990). Outcome in manic disorders. A naturalistic follow-up study. *Archives of General Psychiatry*, **47**, 665–71.

Hooley, J.M. and Teasdale, J.D. (1989). Predictors of relapse in depressives: expressed emotion, marital distress, and perceived criticism. *Journal of Abnormal Psychology*, **98**, 229–35.

Jacobson, N.S., Dobson, K., Fruzzetti, A.E. *et al.* (1991). Marital therapy as a treatment for depression. *Journal of Consulting and Clinical Psychology*, **59**, 547–57.

Joyce, P.R. (1985). Illness behaviour and rehospitalization in bipolar affective disorder. *Psychological Medicine*, **15**, 521–25.

Keitner, G.I. and Miller, I.W. (1990). Family functioning and major depression: an overview. *American Journal of Psychiatry*, **147**, 1128–37.

Keitner, G.I., Ryan, C.E., Miller, I.W. *et al.* (1990). Family functioning, social adjustment, and recurrence of suicidality. *Psychiatry*, **53**, 17–30.

Keitner, G.I., Fodor, J., Ryan, C.E. *et al.* (1991). A cross-cultural study of major depression and family functioning. *Canadian Journal of Psychiatry*, **36**, 254–9.

Kendler, K.S., Neale, M.C., Kessler, R.C. *et al.* (1992). A population based twin study of major depression in women. *Archives of General Psychiatry*, **49**, 257–66.

Kiloh, L.G., Andrews, G., and Neilson, N. (1988). The long-term outcome of depressive illness. *British Journal of Psychiatry*, **153**, 752–7.

Kumar, R. and Brockington, I.F. (1988). *Motherhood and mental illness 2.* Wright, London.

Lee, A.S. and Murray, R.M. (1988). The long-term outcome of Maudsley depressives. *British Journal of Psychiatry*, **153**, 741–51.

McLeod, J.D., Kessler, R.C., and Landis, K. (1992). Speed of recovery from major depressive episodes in a community sample of married men and women. *Journal of Abnormal Psychology*, **101**, 277–86.

Miklowitz, D.J., Goldstein, M.J., Nuechterlein, K.H. *et al.* (1988). Family factors and the course of bipolar disorder. *Archives of General Psychiatry*, **45**, 225–31.

O'Leary, K.D. and Beach, S.R. (1990). Marital therapy: A viable treatment for marital discord. *American Journal of Psychiatry*, **147**, 183–6.

Rounsaville, B.J., Weissman, M.M., Prusoff, B.A. *et al.* (1979). Process of psychotherapy among depressed women with marital disputes. *American Journal of Orthopsychiatry*, **49**, 505–10.

Stephens, J.H. and McHugh, P.R. (1991). Characteristics and long term follow-up of patients hospitalised for mood disorders in the Phipps Clinic, 1913–1940. *Journal of Nervous and Mental Disease* **179**, 64–73.

Stierlin, H., Weber, G., Schmidt, G. *et al.* (1986). Features of families with major affective disorders. *Family Process*, **25**, 325–36.

Weber, G., Simon, F.B., Stierlin, H.S. *et al.* (1988). Therapy for families manifesting manic-depressive behaviour. *Family Process*, **27**, 33–49.

Weissman, M.M., Kidd, K.K., and Prusoff, B.A. (1982). Variability in rates of affective disorder in relatives of depressed and normal probands. *Archives of General Psychiatry*, **39**, 1397–1403.

6

Family aspects of schizophrenia

Campbell (1986), in an extensive review of research on families and health, claims that work on schizophrenia is 10 to 20 years ahead of work on any other disorder. A striking feature has been the increasing influence of empirically based findings, especially those related to the Expressed Emotion concept, and the stimulus provided by these for new family based interventions.

It is noteworthy that family aspects of schizophrenia have assumed special significance at a time when patients with this illness are more likely to live at home than they were in the past. Indeed, families have now clearly become the primary care-givers.

HISTORICAL

An interest in family aspects of schizophrenia arose in the post-World-War-II period. Initially, the influence of psychoanalytical theory was strong. Frieda Fromm-Reichmann developed the notion of the 'schizophrenogenic mother' and this was followed by a number of studies aiming to define interactions pertinent to this supposed attribute. Work in the 1950s and 1960s by researchers such as Lidz made no progress in identifying a specific 'schizophrenogenic' pattern. The focus tended to shift to difficulties in the relationship between the parents of schizophrenic youngsters, and the way in which the sufferer was ensnared in these problems. Bowen at the Family Studies Unit of the National Institute of Mental Health (NIMH) studied the family unit as an interactional whole. Again, marital discord was emphasized while the sufferer was seen as having a role in stabilizing the parental relationship. Also described were problems of individuation in the family, especially due to projective mechanisms and constraining relationship patterns, often transmitted across several generations.

A contribution from an entirely different vantage point was made by Gregory Bateson and co-workers of the Palo Alto group. Their ideas derived from an anthropological approach based on naturalistic observation. Communication patterns in the family, both verbal and non-verbal, were of particular interest. This led to the notion of the 'double-bind' which was postulated to be an important mechanism in the genesis of schizophrenic symptoms.

The subject of the bind receives conflicting communications at verbal and non-verbal levels, but because of his dependent relationship with the message sender, is not able to comment on this inconsistency nor to extricate himself from the relationship. A possible response to this situation, it was suggested, is the development of a communication disorder characteristic of schizophrenia. This group was also prominent in developing the idea of 'family homeostasis', and the ways in which a family may under certain circumstances achieve stability at the expense of the health of its members. Related to this was the notion of the 'solution as problem'. Here an attempted family solution to a problem inadvertently results in a new problem which is sustained by the family's unsuccessful problem-solving behaviour.

While the concept of the 'double-bind' generated excitement it proved extremely difficult to study empirically. Problems were found in its application to real life interactions and little support could be found for a specific association with schizophrenia. The concepts of homeostasis and problem-solving in families continue to be prominent in family therapy approaches, particularly systemic ones.

The 1960s saw many observational studies of families with a schizophrenic member, but few consistent findings emerged. Features such as family efficiency, flexibility, conflict, dominance, coalitions, and distorted communications were studied. Variations in the overall framework of the research, the kinds of observations made, the measures used, and the definitions employed made interpretation of the results difficult. This was exemplified in a review by Doane (1978a), the response by Jacob and Grounds (1978) and the rejoinder by Doane (1978b) to which the interested reader is referred.

In the 1970s research began to focus on two areas which have continued to be the subject of interest: Communication Deviance and Expressed Emotion. We will discuss these after a brief review of genetic factors in schizophrenia.

GENETIC FACTORS

A familial aggregation of schizophrenia clearly occurs and the research data support a genetic component in this. About 30 per cent of patients with schizophrenia have a family history of the disorder. The risk of a child developing schizophrenia if a parent has the condition is approximately 10 per cent. If both parents are affected, it rises to 40–50 per cent. The risk of the child developing schizophrenia is the same regardless of whether the affected parent is the mother or father. The more family members affected, the greater is the risk of a youngster developing schizophrenia, and the more severe the illness is likely to be.

Twin studies are difficult to interpret in terms of their ability to separate genetic from other familial influences. The concordance rate for schizophrenia in monozygotic twins has been found to be consistently higher than that in dizygotic twins (around 50 per cent versus 10–15 per cent). The strongest evidence for a genetic contribution to schizophrenia derives from the NIMH supported Danish adoption studies. The prevalence of schizophrenia and schizophrenia-like disorders was found to be twice as great in the adopted-away offspring of mothers with schizophrenia than in the adopted-away offspring of normal mothers. In an alternative design, diagnoses in the 'schizophrenia spectrum', later reanalysed in terms of DSM-III categories schizophrenia and schizotypal personality, were more common in the biological relatives of adoptees who developed schizophrenia than in relatives of adoptees who did not develop the disorder. A significant excess of schizophrenia in paternal half-siblings of adoptees with schizophrenia argues against a sole influence from intra-uterine factors or early mothering. In another study, biological parents of schizophrenic adoptees showed an approximately five-fold increase in the incidence of schizophrenia compared with the adopting parents of sufferers.

GENE-ENVIRONMENT INTERACTIONS

A Finnish adoption study, still in progress, has produced preliminary findings pointing to an important interaction between genetic and environmental influences (Tienari 1991). 179 adoptees with biological mothers who had schizophrenia were compared with 144 adoptees born of mothers without the illness. All were adopted by non-relative families, the majority before the age of two years. The adoptees in both groups, most of whom were now older than 20, received a careful psychiatric assessment. In addition, the adoptive families were rated for disturbance based on levels of anxiety, boundary functions, parental coalition, quality of interaction, flexibility, conflict, and reality testing. Results to date show that offspring from mothers with schizophrenia are significantly more likely to show severe disturbance ('psychosis' or 'borderline psychosis') than offspring from the control mothers. When disorders in the offspring were related to levels of family disturbance, it was found that disturbance in the offspring increased with increasing family disturbance in both groups. However, the differences between the offspring of mothers with schizophrenia and the offspring of controls only became clear when the adoptive family showed significant disturbance. Forty six per cent of the adoptees of mothers with schizophrenia compared to 24 per cent of the control adoptees showed a severe disturbance ('psychosis', 'borderline psychosis', 'severe personality disorder') when the family was rated as disturbed, but only 5 per cent versus 3 per cent when the family was rated as well-functioning. That is, the youngster tends to develop schizophrenia or a 'borderline schizophrenic' disorder only when his adoptive family shows serious disturbance. This study thus points to an interaction between genetic vulnerability and family disturbance. A poorly functioning family enhances the expression of the genetic vulnerability while a healthy family may protect against such a development. A potential problem with this conclusion concerns the possible effect on family functioning of the disturbed youngster, but as this study also has a prospective component, this issue should be resolvable in the future.

COMMUNICATION DEVIANCE (CD)

This concept was developed by Wynne and Singer (1963*a*,*b*) and refers to a subject's fragmented and disorganized communication patterns, with an inability to maintain a focus. They hypothesized that such communication styles could result in disturbed information processing and thinking in vulnerable offspring. Measures of CD were developed on the basis of the way in which a parent interpreted a Rorschach Test. This group of researchers found that they could differentiate between offspring with and without schizophrenia solely on the basis of parental CD. An important replication of this work was attempted by Hirsch and Leff (1975). They compared CD in relatives of patients suffering from schizophrenia with relatives of neurotic controls. CD was increased in the former group, but this was accounted for by their increased verbosity compared with the control parents. Hirsch and Leff concluded that the relationship between CD and schizophrenia was not supported, but the possibility remains that the increased verbosity may result from greater communication difficulties. It is also possible that CD is a genetic trait which manifests in parents and offspring in the absence of any transactional component, or that CD may be a consequence of living with a youngster with schizophrenia.

In an unusual study, Goldstein (1985) examined family factors that ante-dated the onset of schizophrenia and related disorders. Fifty teenagers who had been seen at a psychological clinic more than 15 years previously were followed-up and a diagnosis determined. At their original presentation, assessments had been made of communication deviance, affective style (AS) (see below), and expressed emotion (EE). By the time of follow-up a sufficient number of youngsters had developed schizophrenia or a schizophrenia-related disorder to allow an examination of the predictive abilities of the three family measures. Despite the small numbers, it was found that 'schizophrenia-spectrum' disorders were significantly associated with a combination of high CD, negative AS, and high EE. Thus CD when associated with negative affective style antedated schizophrenia-like psychopathology. A moderate degree of specificity was suggested because these family qualities related more closely to schizophrenia than to other psychiatric disorders. There are a number of possible interpretations of these results, some of which do not require a causal role for family transactions. This unique study needs replication.

EXPRESSED EMOTION (EE)

This influential concept arose out of studies investigating the effect of family emotional climate on the course of schizophrenia. Early work examined the role of both positive and negative features but later research generally narrowed the concept to negative or intrusive attitudes expressed by relatives about the patient.

EE is assessed on the basis of a semi-structured, standardized interview, the Camberwell Family Interview (CFI), carried out with a relative, singly. Ratings are made from audio-tapes of the interview and are based on both content and vocal tone. The key ratings are (1) critical comments (CC) (a frequency count of statements of resentment, disapproval, or dislike, together with comments showing critical tone irrespective of content); (2) hostility (criticism for what the person *is* rather than for what he *does*, or excessive generalized criticism); and (3) emotional over-involvement (EOI) (unusually marked concern about the patient such as constant worry over minor matters, over-protective attitudes, or intrusive behaviour). Warmth and positive comments are also rated. A designation of 'high' EE is based on a score above a designated threshold level for CC, or hostility, or EOI. Some excellent reviews of EE and its implications have appeared (Kuipers and Bebbington 1988; Kavanagh 1992; Lefley 1992)

Most studies involving EE have focused on its prediction of relapse. Over 20 studies have examined this and the majority have found that high EE is associated with a two- to three-fold increase in relapse rates at 9–12 months follow-up compared with low EE settings. Relapse rates are approximately 50 per cent versus 20 per cent. Naturalistic studies indicate that the effects of high EE may be moderated by less face-to-face contact between relative and patient, and by neuroleptic medication. The role of contact is controversial since this may be confounded by the patient's level of functioning and the relative's social network. While medication moderates the effect of EE it does not remove it. Where EE has been assessed for both parents, it seems that a low EE parent is unable to significantly temper the effect of a high EE parent. Some studies have suggested that the effect of EE on relapse is stronger for males than for females.

Many studies have their methodological problems and there has been a tendency for the more recent to have a greater frequency of negative results for the association between EE and relapse. An example of one of these, which also contains a good review of other studies, is that of Stirling *et al.* (1991). This research involved early onset schizophrenia, all subjects having an illness of less than two years duration. Critical comments were low overall and most high EE ratings were on account of EOI. There was no association with relapse. CC was correlated with failures in pre-illness work/study, the duration of illness, and inversely with the acuteness of onset. It is suggested that EE evolves over time and that there may be a 'reactive' as well as a 'trait' component, especially for criticism. It is possible that EE may be less effective in predicting relapse early in the course of the illness than later.

It is easy to fall into the trap of seeing EE as a static quality of an individual, whereas it is probably better conceptualized as a 'snapshot' of an aspect of the relationship between a relative and patient at a particular point in time. The majority of studies have assessed EE at the time of the patient's admission to hospital and then examined its ability to predict relapse over the next year or so. Assessing EE at admission introduces a possible 'reactive' contribution related to facing a crisis and the effectiveness with which it has been handled, in addition to any ongoing aspect of the relationship it may embody. In one study in which EE was assessed in stable out-patients, it failed to predict relapse over the following 12 months, but over 5 years, there was a tendency for more relapses in families with consistently high EE.

It is difficult to disentangle EE from the duration or severity of the illness, or from aspects of the patient's premorbid and current levels of social functioning. EOI, in particular, may be sensitive to a poorer premorbid adjustment and may increase as the disorder becomes more chronic. Some, but not all, studies have found that critical comments are higher when patients' symptoms are worse and their behaviour more disturbed (for example, MacMillan *et al.* 1986).

The finding that EE changes over time is consistent with a reactive component. When relatives, even without intervention, have been retested 6 to 12 months after the patient's discharge from hospital, 15–45 per cent who were initially high EE become low EE. An initial low EE rating is more stable with only about 15 per cent making the transition to high EE. Critical comments may be more liable to change than EOI but this has not been a consistent finding.

Critical comments and hostility are usually correlated but this is not so for the relationship between critical comments and EOI. While EE is defined as being high on the basis of high criticism or high EOI, the former seems to have made the greatest contribution to relapse in most studies. Cultural factors may be important here as there is evidence that critical comments in some cultures (for example, India) are lower than in England or the United States, while EOI may be higher in some cultures (for example, Portuguese-speaking Brazilians, or Greek families) than in others. It has been suggested that low EE in some cultures may be related to living in extended families which are more tolerant of eccentric behaviour and social withdrawal. The way in which the components of the EE measure relate to each other may also vary across cultures.

There is a claim that people rated high in EE are less informed about schizophrenia and are more likely to attribute difficult behaviour to personal characteristics of the patient rather than to the illness. Negative symptoms may be especially prone to such attributions. Brewin *et al.* (1991) found that relatives scoring high on criticism made more personal attributions than low EE relatives or those high only on EOI. Hostility was associated with behaviour being seen as wholly due to the patient with few external causal elements. There is also evidence that high EE families are more likely to be unpredictable in their responses to the patient. Low EE relatives are often described as calm, empathic, and respectful of the patient's often peculiar needs.

Mechanism of relapse

EE is usually seen as relating to relapse by raising the patient's arousal beyond an optimal level, thus increasing the likelihood of the expression of a biological vulnerability to develop schizophrenic symptoms. In this 'stress-diathesis' model, arousal and an associated impairment in the capacity for information processing play a crucial role. This conceptualization has led to a number of studies examining the effects of EE on the patient's psychophysiology (see Tarrier 1989). There is experimental support for a greater tonic frequency of spontaneous fluctuations in skin conductance (SC) in patients from high EE homes, especially following a significant life event. Other studies have shown a faster rate of habituation from elevated levels in the presence of a low EE relative compared to a high one. However, this difference may not be consistent over time as one study found that such differences were not evident nine months after discharge from hospital, with patients in both groups showing a decrease in SC. Despite these findings, SC responses did not predict relapse.

A simple model in which high EE causes relapse in a vulnerable patient, or alternatively, in which EE is seen as an epiphenomenon of the behavioural disturbance of schizophrenia, is unlikely to be satisfactory.

Reciprocal interactions between the patient and relatives make the processes more complex. It has been argued, for example, that some problem behaviours may lead to unsuccessful attempts at coping by relatives which may lead to the expression of high EE, which in turn may exacerbate the problem behaviours and symptoms. The types of strategies used by relatives may be important here, with coercive measures being more likely associated with high EE and patient withdrawal. A study, so far unpublished, by Smith and colleagues in Birmingham has shown that high EE relatives experience more distress and burden, and perceive themselves as coping less well than low EE relatives. However, approximately one quarter of the low EE relatives reported high stress, burden, or impaired coping. These workers have also raised the possibility that in some low EE families there may be detachment or even neglect, reducing stress levels but at the expense of fostering apathy, negative symptoms, and social impairment in the patient. A developmental and interactional view of EE is begining to emerge.

EE and family interactions

Despite the value of EE in predicting relapse very little is known about what it means in terms of family interaction. The few studies to date show fairly consistently that high EE parents are more critical in direct interaction with their ill offspring and that negative patterns of interaction arise between them.

One approach has examined measures of Affective Style (AS) in which transcripts of direct interactions are rated for critical or intrusive statements. High EE critical parents are more likely to show the former, while high EOI is associated with the latter. The patient's contribution to the interaction has also been examined. Hahlweg *et al.* (1989) examined both verbal and non-verbal behaviour on a unit by unit basis. Interactions between a parent and patient were then studied sequentially. High EE critical parents, when discussing emotion-laden family problems, showed a negative interactional style with more criticism, more negative non-verbal affect, and more negative solution proposals. Low EE parents showed more positive and supportive statements. Patients with critical relatives also showed more negative non-verbal affect, expressed more disagreement, and made more self-justifying statements. Sequential analyses showed longer sustained negative reciprocal patterns, especially non-verbal, in critical families.

Strachan *et al.* (1989) also examined patients' coping styles during family interactions with high or low EE relatives. The results were somewhat complex but patients were more critical in interaction with a critical parent, and were more likely to make 'autonomous' statements (indicating a desire to follow a self-motivated path) with low EE parents. The patient's contribution to the interaction was most negative when both parents were rated high EE. Self-denigratory statements were also more common when patients interacted with high EE parents. Symmetrical (criticism) and complementary (self-denigratory) patterns of transaction were thus evident. This is consistent with another study involving non-schizophrenic adolescents in which a reciprocity was demonstrated between high or low EE mothers and the degree of warmth or coldness shown by their offspring. High EE mother-child dyads were also more tightly 'joined' and less flexible in their interaction than the low EE dyads.

In another approach, Hubschmid and Zemp (1989) conducted detailed semi-structured interviews with parents, independently assessed on EE, concerning interactions between them and the patient. High EE was associated with interactions having a negative affective tone, conflict, and rigid patterns. Parents scoring high on EE made fewer 'helping and protecting' and more 'belittling and blaming' responses. EOI was associated with more 'ambiguous statements', implying a special knowledge or insight into their offspring's feelings and thoughts.

Blame

The EE concept has not proved popular with families, nor with relatives' organizations, being sometimes perceived as yet another means of blaming relatives for the patient's illness. Kavanagh (1992) cautions against the term EE being used as an adjective for people or families. It is better seen not as an enduring characteristic of a person but as a measure of the 'emotional temperature' of an environment which may change over time. He also notes that the literature tends to focus on negative aspects of family interactions without recognizing the many families who have managed to cope successfully with a sick member over long periods without developing negative attitudes. He suggests that it is important to acknowledge families for finding successful ways of dealing with difficult problems. Indeed, most families are found to be low EE. A focus on the protective aspects of low EE has been urged by some as a counterbalance to the more common emphasis on determining what is detrimental about high EE (Lefley 1992).

PSYCHOSOCIAL INTERVENTIONS

A number of family based interventions for schizophrenia, largely stimulated by the work on EE, have emerged. These have focused on patients in high EE settings who are thus considered to be at a high risk of relapse.

The interventions have been offered as part of a treatment package which includes medication and routine psychiatric care. The aims have been to prevent relapse following discharge from hospital (perhaps by changing EE from high to low) and, to a lesser degree, to improve the patient's level of functioning and increase family well-being. This area has been well reviewed by Lam (1991).

Education

Several investigations have examined the effectiveness of education alone. This has usually involved one to six sessions with individual families or in groups, with the patient usually absent. Written information is also sometimes provided, and in one study was the only means of transmission. The information presented has usually covered the nature of schizophrenia, the symptomatology and its relationship to behavioural disturbances, aetiology, course and outcome, and the rationale for medication and other treatments. Some programmes have also included information on ways of coping with difficult behavioural problems, or information about professional resources available. It has been hoped that a gain in such knowledge will also be effective in changing relatives' attitudes and beliefs about the disorder including inaccurate attributions concerning the patient's behaviour (for example, negative symptoms being misconstrued as 'laziness').

These educational interventions have had a significant but modest impact. An increase in knowledge has been demonstrated which has been maintained over six months. There has been some reduction in family distress but much of this may be transient. An increase in optimism concerning the family's role in maintaining the patient's well-being has been noted, but not an effect on reducing relapse. Effects on criticism have been minor. The incorporation of material concerning coping methods and problem-solving approaches may lead to a better outcome, but the long-term effectiveness is unknown. In one study a correlation was found between the acquisition of knowledge and the other beneficial effects, suggesting that the information provided, rather than a 'placebo' effect', was the important factor in the improvement (Birchwood *et al.* 1992). Significant differences between different methods of promoting knowledge have not been demonstrated. For example, in the study mentioned above, postal information, videotapes, and relatives' groups were equally effective at six months follow-up, although the last was superior earlier on. This study also found that relatives who were more distressed and who perceived the patient as having greater control over his or her symptoms were more likely to drop-out. Many believe that presenting information to families individually, tailoring this to their special needs, and taking account of their own 'explanatory models' is likely to prove superior, but this has not been tested.

The timing of the offer for help may be important. Families may be more likely to accept when they have experienced a recent crisis. Different effects may occur at different phases of the illness. While it might be thought that information should be maximally valuable early in the course of the illness, families may be less receptive at this stage because of uncertainty or fear regarding the prognosis.

An educational component has been an integral part of the family interventions to be discussed below where, in addition to any other benefits, it has been regarded as helpful in engaging families.

Family interventions

To date there have been seven family intervention studies with careful outcome evaluations. Five of these examined outcome at 9–12 months and again at two years (Leff *et al*. 1985, 1988, 1990, 1992; Falloon *et al*. 1985, 1987; Hogarty *et al*. 1986, 1991; Tarrier *et al*. 1988, 1989). Two examined outcome only at 9–12 months (Kottger *et al*. 1984; Vaughan *et al*. 1992). The major focus of these studies was a reduction in relapse rates but some also examined the patient's social functioning, levels of family distress, and changes in EE. Most of the interventions had common components but there was some variation. The majority treated the patient with the family, but relatives' groups were also used. Apart from education, the interventions usually offered training in the management of the patient's illness and difficult behaviours, support, measures to build morale, strategies for strengthening family organization and enhancing stability or predictability, and attempts to change relatives' attitudes so decreasing high EE. In these studies there were around 12 to 25 patients in each treatment condition.

Five studies showed a significantly better outcome for family interventions compared with standard care. Overall the relapse rate with the family intervention at nine months was around 10 per cent compared with 50 per cent for routine or individual treatment. At two years it was 33 per cent compared with 71 per cent. The best outcome at two years occurred with those interventions that were maintained for the full two year period (Falloon 17 per cent; Hogarty 25 per cent). In the study of Falloon *et al*. (1985, 1987) the family treatment group had patients functioning socially at a higher level, a better level of family social adjustment, and a greater improvement in relatives' burden and distress. In the study of Hogarty *et al*. (1986) such improvements in patient and family functioning were less evident. Where it was examined, relatives in the family treatment group showed a significantly greater reduction in EE than in the control treatment group. However, it cannot be concluded that the reduction in EE was the major cause of the better outcome in family treatment since a change from high to low EE was not necessary for a good outcome.

The effectiveness of the interventions did not seem to be due to better drug compliance; in the Falloon *et al.* study, there was a trend for lower doses in the family treatment condition. Hogarty *et al.* found that a combination of social skills training for the patient together with family treatment resulted in a better outcome at nine months (but not at two years) than with family treatment alone. A contribution to the better outcomes from the extra attention received by the families, acting in some non-specific manner, cannot be excluded, but is unlikely to have been major.

The two studies which did not find an improved outcome for the family treatment approach differed significantly from the other five studies. The Hamburg study (Kottgen *et al.* 1984) adopted a group psychoanalytical approach in which the patient and relatives were seen separately. The Sydney trial (Vaughan *et al.* 1992) also intervened with the relatives alone. The intervention was behaviourally orientated but was much briefer (10 weekly sessions) than in the successful interventions discussed above. There was also less involvement by the therapist in the general management of the patients.

While there were differences between the successful therapeutic approaches there were also some common components. Lam (1991) has attempted to identify these:

1. A positive approach to the family and the establishment of a genuine working relationship. Positive aspects of the family, and the family's ability to change were underlined; members' own needs were respected.

2. The provison of structure and stability. Attempts were made to set limits on the patient's behaviour while maintaining a 'moderate' interpersonal distance.

3. A focus on the 'here and now'. The family's coping strategies were examined and strengths and weaknesses noted. Current perceptions rather than their history were the main interest.

4. The use of family concepts. Attention to interpersonal boundaries and the marital coalition seem to have been a general concern.

5. Cognitive restructuring. An educational thrust which attempted to reduce guilt and to change unhelpful attributions for problematic behaviours.

6. A behavioural approach. A behavioural analysis of needs and strengths, with a setting of realistic goals, defining small steps towards their attainment, task assignment, and monitoring of change. In some interventions this was presented within a 'problem-solving' framework.

7. Improving communication. This generally aimed for clarity, directness, and specificity, usually within the context of problem-solving.

We will consider the implications of these studies for routine practice later.

FAMILY 'BURDEN'

The impact on the family of a patient with schizophrenia has been comparatively neglected, despite the fact that with increasing deinstitutionalization relatives have become the primary carers. About 50 per cent of patients with schizophrenia live with a supporter, usually a relative, and most commonly parents for males and a husband for females. A single parent, usually a mother, becomes increasingly common as patients become older (Gibbons *et al.* 1984).

Relatives complain uncommonly to professionals about their difficulties. On the other hand, when surveyed directly, most express dissatisfaction with the support received (Creer and Wing 1974). They feel insufficiently informed about the illness and ways of managing it, and inadequately consulted in medical decisions concerning their ill family member. In the Gibbons *et al.* study, about 70 per cent of caregivers reported symptoms of emotional or physical ill-health and a third had scores on the General Health Questionnaire above threshold for psychiatric cases. About 50 per cent reported moderate to severe distress as a result of the patient's behaviour. This was highest during the first year of the patient's illness. There was a significant correlation between relatives' distress and the level of the patient's symptomatic disturbance.

The notion of burden is a complex one and, being so closely tied to expectations (often idiosyncratic), and performance, is difficult to measure in any absolute sense. Attempts have been made to separate burden into two components – 'objective' (for example, loss of earnings, extra household tasks, restriction of social activities) and 'subjective' (the degree of subjective distress experienced). These are not highly correlated. Such a conceptualization may miss significant aspects of the experience of care-giving which cause distress. These include grief for the person the patient once was or showed promise of becoming, stigma and its effects on family and outside relationships, guilt about having transmitted the disorder or about not having done enough to prevent it, and frustrations with psychiatric services.

Creer and Wing (1974) found that two groups of behaviours were experienced as most burdensome. The first was related to 'negative symptoms' – social withdrawal, underactivity, slowness, and lack of interest or conversation. These are particularly difficult to tolerate for many relatives, especially when they become the patient's sole social contact. Disturbed, socially embarrassing, or unpredictable behaviour was the second. Problems with violence or suicidal intent are less common but dominate when they occur.

In a study being carried out by two of the authors (GS, SB) in Melbourne, dimensions of burden are evident which rarely figure in conventional accounts. Some of these relate to the wider family system; many families face additional ill-health in other family members (including schizophrenia), other losses (for example, a suicide), or a recrudescence in a new guise of old conflicts (for example, when parents-in-law are critical of a caregiver wife for not looking after their schizophrenic son adequately and fail to provide support).

Kuipers and Bebbington (1985) have suggested ways in which burden may be linked to relatives' ways of coping with the illness, EE, and illness outcomes, but there have been no direct evaluations of the relationship between these variables. It is possible that relatives who experience greater distress are those who have less effective coping styles and who score highly on EE. There is some evidence that low EE is associated with ways of dealing with problem behaviours that are centred on the patient, while high EE may be associated with more concerns about the relative's own responses to the situation.

Birchwood and Cochrane (1990) examined family coping styles in relation to patients' behavioural problems and relatives' distress. They found a consistency of coping style across categories of problem behaviours. A number of styles were identified based on relatives' verbal responses to videotapes of behavioural problems. A 'coercive' style (using punitive or critical approaches) was adopted by parents with low social functioning patients, especially social withdrawal. A 'disorganized' style (desperate and inconsistent strategies) correlated with more severe levels of patient disturbance. An 'ignore/accept' style (being acceptance or a deliberate strategy of non-response) was associated with lower burden, but when the contribution of behavioural disturbance was 'partialled out', this effect disappeared. A 'coercive' style related to less sense of being in control while an 'ignore/accept' style was associated with the opposite. This study highlights the need for an interactional perspective in attempting to understand the relationships between burden and behaviour.

We do not know yet how best to help relatives with these difficulties. Educational interventions may have an effect on relatives' distress or burden (see above). Some of the psychosocial interventions aimed at preventing relapse have also examined their effects on burden. This was significantly reduced in Falloon *et al.*'s study, but less consistently so in Hogarty *et al.*'s. Kuipers and Bebbington suggest a number of practical approaches which will be considered later.

OTHER FAMILY THERAPY APPROACHES

Psychoanalytically orientated family therapy approaches have come under fire in the past and are no longer commonly used.

Systemic and strategic family therapies have been used in schizophrenia, with the accounts of Selvini-Palozzoli and Haley being especially influential. Jones (1987) has presented an accessible account of the underlying assumptions and techniques of the systemic approach. She expresses a view in sharp contrast with that espoused by practitioners of the psychosocial treatments described above: 'I will continue to believe that I can serve clients better by working with them according to the premise that they may be able to engage fully in life, having gained strength and flexibility from successfully negotiating potentially disastrous experiences'. The implication is that a 'cure' is the goal, not an acceptance of impairments needing containment. The symptoms of the illness are seen as being maintained by, and as maintaining, in a circular manner, a dysfunctional pattern of family interactions. The aim of therapy is to disrupt the pattern in such a way as to make it difficult for the family to continue to behave as before, and to permit them to find a more flexible organization in which the illness has no place. The techniques of circular questioning, neutrality, reframing, positive connotation, prescribing the symptom, and intersession tasks are used. Jones claims 'good results' from her work in 90–95 per cent of patients with an illness of less than 5-years duration. However, no data on outcome which meet minimum scientific standards have been presented by any practitioners using this approach. Thus their claims cannot be evaluated.

Strategic family therapy as described by Haley takes a pragmatic form, concentrating on the structure of the family, particularly breakdowns in its hierarchy. Parents are often seen as having surrendered control so that clashes with the sick youngster tend to end inconclusively. Strategies have been developed which help the parents regain control, in the process enabling the patient to progress with the developmental task of separating from them. Again outcome data are not available, but several of these techniques have found a place in the psychoeducational interventions mentioned earlier.

IMPLICATIONS FOR EVERYDAY PRACTICE

Despite the impressive body of research discussed in this chapter, difficulties remain in assessing its relevance for the general psychiatrist. An important role for families in management has been demonstrated, but how are the research findings to be translated into clinical practice? If a new drug were discovered to have similar efficacy in reducing relapse there seems little doubt that it would be widely used by now. But, to date there has been little impact. Why is this so?

One problem concerns the generalizability of the findings from the psychosocial interventions. All targeted high EE relatives, so ascertained at the point of admission to hospital. The assessment of EE is a complex and time-consuming procedure, not feasible in ordinary practice. Furthermore, it is not known whether low EE families might not also benefit from such treatment; relapse rates are still substantial, while there is likely to be scope for improvement in social functioning and in the burden experienced by relatives.

Perhaps even more important are certain practical issues. One is the failure to engage a substantial number of apparently suitable families; dropout rates may be up to 35 per cent. An impression exists that many families among the most distressed or poorly coping are not disposed to accept the offer of family intervention. Leff and colleagues found in one intervention study that relatives who failed to attend even once, changed the least, while the patients fared the worst. Shame, guilt, hopeless resignation, denial, or anger with the treatment so far received may be important here. Being offered a family treatment may suggest to the family that they are considered a 'problem family' rather than a family with a problem. A second issue concerns the level of skill required to be effective. It is always difficult to know how a psychological treatment expertly administered by an enthusiastic research team will fare when it becomes a routine measure in the hands of regular staff. Other questions arise from some of the specifics of the interventions. While similarities between the approaches adopted can be abstracted as we have seen, differences remain. For example, the emphasis on behavioural methods varies, as does the balance between home-based versus clinic-based sessions.

Finally, the effectiveness of family interventions in the long-term has been questioned. The notion of relapse prevention has been replaced with one of relapse postponement or delay. While a significant difference remains between experimental and control groups at 2-year follow-up, patients continue to relapse at a substantial rate in the second year despite treatment. Hogarty *et al.* (1991) sound a pessimistic note in their report. A post-study evaluation found a relapse rate of 80 per cent in all groups. Patients remained vulnerable throughout and many relapsed when greater performance demands were made of them in the later phases of treatment. In this study at least, patients and families did not seem to acquire skills exercised autonomously which could forestall future relapses. The effect of therapy was likened by these authors to those of prophylactic medication; when discontinued, the risk of relapse remained high.

Some recent reports have examined ways of incorporating the fruits of research into routine practice.

In Nithsdale, a 'package' of treatment comprising educational seminars, relatives' groups, and family meetings in the home was offered to the relatives of all patients with schizophrenia living at home. The staff were not specially trained but were experienced in psychiatric care. Fifty per cent of relatives did not accept the offer from the outset, while the remainder attended variably. Reasons for refusing the intervention included 'things are fine at the moment', 'it's the patient who needs help, not me', 'the patient does not want anyone else to know that he has been ill', 'too busy', 'I've coped all these years', and 'it's too late now'. Acceptance was more likely if there had been a recent admission to hospital. For those who did participate there was a trend to a reduction in relapses over the next 18 months compared to the pre-intervention 18 months. The relatives were also reasonably pleased with the treatment. In most cases EE did not change over the period of the intervention.

Brooker *et al.* (1992). reported a study in which community psychiatric nurses (CPNs) were trained in the delivery of family interventions. Outcomes for a group of patients treated in the community by these CPNs were compared with a control group treated by CPNs in a routine manner. Since the patients could not be randomized, the results must be viewed with caution. Important undetected differences between the two groups might have existed prior to the intervention, and a selection bias may have operated in the choice of the CPNs who were trained. Nevertheless, the results were promising. The experimental group showed improvement in some target symptoms (anxiety, depression, and retardation) and there was a tendency for a reduction of neuroleptic dosage. An improvement in social functioning of the patients was noted, as was a diminution in minor psychiatric morbidity in the relatives. Consumer satisfaction (information, practical advice, emotional support, service coordination, attitudes of professionals) also increased significantly. These changes did not occur in the control group.

In Birmingham, a programme has been designed which promotes and integrates family intervention within a general service (Smith and Birchwood 1990). The principles have been carefully thought out by the instigators. Admission EE is not used to select suitable families, since it is argued that patients who are not admitted may still benefit from treatment as may low EE families, particularly if the focus is shifted from preventing relapse to improving social functioning and family well-being. Difficulties in engaging families due to EE's perceived pejorative connotation should also thus diminish. A needs-led, goal-defined intervention is advocated. Since EE is not very informative about what actually is happening in the family, its omission, it is argued, poses no real handicap. The programme aims actively to engage patients through a multiplicity of contact points in the psychiatric services, to ensure that a particular family's needs are understood and responded to within an 'informed partnership' with the family, and to negotiate the goals of the intervention with the family. Provision for training and supervision of 'front-line' professionals is a key aspect, together with means of monitoring the services to ensure that such dissemination of skills does not prejudice the quality of the treatment.

At the Maudsley Hospital in London, education and a monthly relatives' group, carried out by the clinical team, were offered to relatives of long-term mentally ill patients (McCarthy *et al.* 1989); a quarter declined. For those that participated, a reduction in EE resulted, together with improved coping skills compared with a control group. Relapse rates were no different, but patients in the experimental group showed a significant improvement in social functioning.

We thus see early attempts to translate research findings into programmes for everyday practice. Problems with engagement loom large, but there are indications that interventions remain effective, especially if the focus is broadened beyond simple relapse. Further reports from these as well as other centres are awaited with interest.

What are the implications for the clinician in terms of his personal practice? We draw a number of conclusions, some necessarily tentative.

Assessment

The stress-diathesis model is helpful in orienting the clinician to important family parameters, and indicates how stressors within the family, as well as its capacity to buffer the patient from those outside may be significant.

Several areas of particular importance in the general assessment of the family (Chapter 3) can be highlighted. These include the way in which the illness has affected the family, and family members' understanding of its nature and treatment. The attributions placed on symptoms, particularly 'negative' ones, should be ascertained as part of an exploration of the family's idiosyncratic 'model' of the illness. A functional analysis of the family from a behavioural viewpoint may be useful, examining especially the clarity of communication, coping style, the contingencies surrounding problem behaviours which may serve to sustain them, and problem-solving skills. In this perspective the delineation of strengths as well as weakness is underscored. Structural issues such as boundaries and optimal interpersonal distances should also be examined. Transgenerational patterns and family rules do not figure prominently in the psychosocial intervention literature, but it is likely that they influence the way in which the family responds to both illness and interventions. The clinician may draw conclusions about whether a high EE atmosphere prevails, but it is probably more useful to approach the assessment through an examination of the interaction between problem behaviours and coping responses since this is linked more directly with treatment recommendations. The family's sensitivity to blame should be particularly examined, including subtle, and sometimes not so subtle, messages received from professionals in the past.

Management

Given the evidence, a positive approach to the family and the provision of information about the illness are important aims. Most families want to know more and discussion along these lines is helpful in building a partnership and opening the door to shared decision-making. Care especially is required when the episode is the first. The prognosis may be uncertain and needs to be discussed in terms of the range of possible outcomes. Education may go further than communicating facts about schizophrenia. The rationale for specific treatments, help in recognizing the warning signs of relapse, and the ability to discriminate between symptoms and drug side-effects might also be considered. Tarrier and Barrowclough (1986) provide useful guidelines, based on general research in the field, for communicating information as part of an interactional process.

Allowing family members to tell their story and acknowledging their care-giving experiences, both negative and positive, is usually much appreciated by them. The question of blame will often need to be addressed explicitly. How much further the clinician is prepared to venture into family involvement will depend on the family's wishes as well as his assessment of his own competence. The principles underlying further treatment have been elucidated by research and provide a framework for planning specific interventions. Kuipers (1991) and Kuipers *et al.* (1992) have given accounts of these which, although cast in an EE context, are aimed at the non-expert. Important measures include a reduction of criticism by changing attributions and improving coping, and reducing emotional over-involvement by determining realistically the degree of independence the patient is capable of achieving, reducing guilt, and by emphasizing the carer's own needs, including 'permission' for these to be satisfied.

The general approach advocated is a problem-solving one, selecting difficulties which are important to the family, setting agreed, specified, and limited goals, and monitoring positive changes and their effects on the family's confidence. Progressing in small steps, with a focus on specific issues (for example, getting out of bed), while setting practical targets within a realistic time-scale for change are important. Knowing when something cannot be changed is equally so. Fostering listening and clear communications may also play a role. A stable, long-term treatment relationship will usually be necessary to achieve these goals. Relatives' groups or self-help groups may also be recommended.

CONCLUSIONS

An important shift in thinking about the family in schizophrenia has occurred since the 1970s. The family as 'cause' has all but disappeared, while a series of excellent studies has tackled the ways in which families may be helped to improve the patient's condition. Interventions are no longer 'all-purpose' family therapies, but are becoming increasingly specific in their targets. Serious attention is also being given to the negative experiences of caregiving. While EE has proven the driving force for much research, there is an increasing shift towards an interactional view of family processes with a wider focus. It is likely that everyday clinical practice will be increasingly informed by these developments.

REFERENCES

Birchwood, M. and Cochrane, R. (1990). Families coping with schizophrenia: coping styles, their origins and correlates. *Psychological Medicine*, **20**, 857–65.

Birchwood, M., Smith, J., and Cochrane, R. (1992). Specific and non-specific effects of educational intervention for families living with schizophrenia: a comparison of three methods. *British Journal of Psychiatry*, **160**, 806–14.

Brewin, C.R., MacCarthy, B., Duda, K., and Vaughn, C.E. (1991). Attribution and expressed emotion in the relatives of patients with schizophrenia. *Journal of Abnormal Psychology*, **100**, 546–54.

Brooker, C., Tarrier, N., Barrowclough, C. *et al.* (1992). Training community psychiatric nurses for psychosocial intervention: report of a pilot study. *British Journal of Psychiatry*, **160**, 836–44.

Brown, G.W., Birley, J.L.T., and Wing, J.L. (1972). Influence of family life on the course of schizophrenia disorders: replication *British Journal of Psychiatry*, **121**, 241–58.

Campbell, T.L. (1986). Family's impact on health: A critical review. *Family Systems Medicine*, **4**, 135–200.

Cozolino, L.J., Goldstein, M.J., Nuechterlein, K.H. *et al.* (1988). The impact of education about schizophrenia on relatives varying in expressed emotion. *Schizophrenia Bulletin*, **14**, 675–87.

Creer, C. and Wing, J.K. (1974). *Schizophrenia at home*. National Schizophrenia Fellowship, Surbiton.

Doane, J.A. (1978*a*). Family interaction and communication deviance in disturbed and normal families: A review of research. *Family Process*, **17**, 357–76.

Doane, J.A. (1978*b*). Questions of strategy: Rejoinder to Jacob and Grounds. *Family Process*, **17**, 389–94.

Doane, J.A., West, K.L., Goldstein, M.J. *et al.* (1981). Parental communication deviance and affective style: predictors of subsequent schizophrenia spectrum disorders in vulnerable adolescents. *Archives of General Psychiatry*, **38**, 679–85.

Doane, J.A., Goldstein, M.J., Miklowitz, D.J., and Falloon, I.R.H. (1986). The impact of individual and family treatment on the affective climate of families of schizophrenics. *British Journal of Psychiatry*, **148**, 279–87.

Fadden, G.B., Bebbington, P.E., and Kuipers, L. (1987). The burden of care: the impact of functional psychiatric illness on the patient's family. *British Journal of Psychiatry*, **150**, 285–92.

Falloon, I.R.H. and Pederson, J. (1985). Family management in the prevention of morbidity of schizophrenia: the adjustment of family unit. *British Journal of Psychiatry*, **147**, 156–63.

Falloon, I.R.H., Boyd, J.L., McGill, C.W. *et al.* (1985). Family management in the prevention of exacerbation of schizophrenia. A controlled study. *New England Journal of Medicine*, **306**, 1437–40.

Falloon, I.R.H., Boyd, J., and McGill, C. (1984). *Family care of schizophrenia*, Guilford Press, New York.

Falloon, I.R.H., Boyd, J.L., McGill, C.W. *et al.* (1982). Family management in the prevention of morbidity of schizophrenia. Clinical outcome of a two-year longitudinal study. *Archives of General Psychiatry*, **42**, 887–96.

Falloon, I.R.H., McGill, C.W., Boyd, J.L., and Pederson, J. (1987). Family management in the prevention of morbidity of schizophrenia: a social outcome of a two-year longitudinal study. *Psychological Medicine*, **17**, 59–66.

Gibbons, J., Horn, S., Powell, J. *et al.* (1984). Schizophrenic patients and their families: a survey in a psychiatric service based on a DGH unit. *British Journal of Psychiatry*, **144**, 70–7.

Goldstein, M.J. (1985). Family factors that antedate the onset of schizophrenia and related disorders: the results of a fifteen year prospective longitudinal study. *Acta Psychiatrica Scandanavica*, **71**, Suppl 319, 7–18.

Goldstein, M.J., Rodnick, E.H., Evans, J.R. *et al.* (1978). Drug and family therapy in the aftercare of acute schizophrenics. *Archives of General Psychiatry*, **35**, 1169–77.

Hahlweg, K., Goldstein, M.J., Nuechterlein K.H. *et al.* (1989). Expressed emotion and patient-relative interaction in families of recent onset schizophrenics. *Journal of Consulting & Clinical Psychology*, **57**, 11–18.

Hirsch, S.R. and Leff, J.P. (1975). *Abnormalities in parents of schizophrenics*. Oxford University Press, Oxford.

Hogarty, G.E., Anderson, C.M., Reiss, D.J. *et al.* (1986). Family psychoeducation, social skills training and maintenance chemotherapy in the aftercare treatment of schizophrenia. 1. One-year effects of a controlled study on relapse and expressed emotion. *Archives of General Psychiatry*, **43**, 633–42.

Hogarty, G.E., Anderson, C.M., Reiss, D.J. *et al.* (1991). Family psychoeducation, social skills training, and maintenance chemotherapy in the aftercare treatment of schizophrenia. *Archives of General Psychiatry*, **48**, 340–7.

Hubschmid, T. and Zemp, M. (1989). Interactions in high and low EE families. *Social Psychiatry and Psychiatric Epidemiology*, **24**, 113–19.

Jacob, T. and Grounds, L. (1978). Confusions and conclusions: A response to Doane. *Family Process*, **17**, 377–87.

Jones, E. (1987). Brief systemic work in psychiatric settings where a family member has been diagnosed as schizophrenic. *Journal of Family Therapy*, **9**, 3–25.

Kavanagh, D.J. (1992). Recent developments in expressed emotion and schizophrenia. *British Journal of Psychiatry*, **160**, 601–20.

Köttgen, C., Sonnichsen, I., Mollenhauser, K., and Jurth, R. (1984). Group therapy with the families of schizophrenic patients: results of the Hamburg Camberwell Family Interview Study. III. *International Journal of Family Psychiatry*, **5**, 84–94.

Kuipers, L. (1991). Schizophrenia and the family. *International Review of Psychiatry*, **3**, 105–17.

Kuipers, L. and Bebbington, P.E. (1985). Relatives as a resource in the management of functional illness. *British Journal of Psychiatry*, **147**, 465–71.

Kuipers, L. and Bebbington, P. (1988). Expressed emotion research in schizophrenia: theoretical and clinical implications. *Psychological Medicine*, **18**, 893–909.

Kuipers L., MacCarthy B., Hurry, J., and Harper, R. (1989). Counselling the relatives of the long term mentally ill. II. A low cost supportive model. *British Journal of Psychiatry*, **154**, 775–82.

Kuipers, L., Leff, J., and Lam, D. (1992). *Schizophrenia family work: a practical guide*. Gaskell, London.

Lam, D.H. (1991). Psychological family intervention in schizophrenia: a review of empirical studies. *Psychological Medicine*, **21**, 423–41.

Leff, J.P., and Vaughn, C. (1985). *Expressed emotion in families*. Guilford Press, New York.

Leff, J.P., Kuipers, L., Berkowitz, R. *et al.* (1982). A controlled trial of social intervention in schizophrenia families. *British Journal of Psychiatry*, **141**, 121–34.

Leff, J.P., Kuipers, L., Berkowitz, R., and Sturgeon, D. (1985). A controlled trial of social intervention in the families of schizophrenia patients: two-year follow-up. *British Journal of Psychiatry*, **146**, 594–600.

Leff, J.P., Berkowitz, R., Shavit, N. *et al.* (1988). A trial of family therapy v. a relatives' group for schizophrenia. *British Journal of Psychiatry*, **153**, 58–66.

Leff, J.P., Berkowitz, R., Shavit, N. *et al.* (1990). A trial of family therapy versus a relatives' group of schizophrenia. Two-year follow-up. *British Journal of Psychiatry*, **157**, 571–7.

Lefley, H.P. (1992). Expressed emotion: Conceptual, clinical, and social policy issues. *Hospital and Community Psychiatry*, **43**, 591–8.

MacCarthy, B., Kuipers L., Hurry, J. *et al.* (1989). Counselling the relatives of the long term mentally ill. I. Evaluation. *British Journal of Psychiatry*, **154**, 768–75.

MacMillan, J.F., Gold, A., Crow, T.J. *et al.* (1986). Expressed emotion and relapse. *British Journal of Psychiatry*, **148**, 133–43.

Miklowitz, D.J., Goldstein, M.J., Falloon, I.R.H. *et al.* (1984). Interactional correlates of expressed emotion in the families of schizophrenics. *British Journal of Psychiatry*, **144**, 482–7.

Smith, J. and Birchwood, M.J. (1987). Specific and non-specific effects of educational intervention with families living with schizophrenic relatives. *British Journal of Psychiatry*, **150**, 645–52.

Smith, J. and Birchwood, M. (1990). Relatives and patients as partners in the management of schizophrenia: the development of service model. *British Journal of Psychiatry*, **156**, 654–60.

Stirling, J., Tantam, D., Thomas, P. *et al.* (1991). Expressed emotion and early onset schizophrenia: a one year follow-up. *Psychological Medicine*, **21**, 675–85.

Strachan, A.M., Feingold, D., Goldstein, M.J. *et al.* (1989). Is expressed emotion an index of a transactional process? II. Patient's coping style. *Family Process*, **28**, 169–81.

Tarrier, N. and Barrowclough, C. (1984). Providing information to relatives about schizophrenia: some comments. *British Journal of Psychiatry*, **149**, 458–63.

Tarrier, N., Barrowclough, C., Porceddu, K., and Watts, S. (1988). The assessment of psychophysiological reactivity to the Expressed Emotion of the relatives of schizophrenic patients. *British Journal of Psychiatry*, **152**, 618–24.

Tarrier, N., Barrowclough, C., Vaughn, C. *et al.* (1988). The community management of schizophrenia: a controlled trial of a behavioural intervention with families to reduce relapse. *British Journal of Psychiatry*, **153**, 532–42.

Tarrier, N., Barrowclough, C., Vaughn, C. *et al.* (1989). Community management of schizophrenia; a two-year follow-up of a behavioural intervention with families. *British Journal of Psychiatry*, **154**, 625–8.

Tarrier, N. (1989). Electrodermal activity, Expressed Emotion in individual and family settings : a comparative study. *British Journal of Psychiatry* (*supplement*) **5**, 51–6.

Tienari, P. (1991). Interaction between genetic vulnerability and family environment: The Finnish adoptive study of schizophrenia. *Acta Psychiatrica Scandanavica*, **84**, 460–5.

Vaughan, K., Doyle, M., McConaghy, N. *et al.* (1992). The Sydney intervention trial: a controlled trial of relatives' counselling to reduce schizophrenic relapse. *Social Psychiatry and Psychiatric Epidemiology*, **27**, 16–21.

Wynne, L. and Singer, M. (1963a). Thought disorder and family relations of schizophrenics. I. Research strategy. *Archives of General Psychiatry*, **9**, 191–8.

Wynne, L. and Singer, M. (1963b). Thought disorder and family relations of schizophrenics. II. A classification of forms thinking. *Archives of General Psychiatry*, **9**, 199–206.

7

Family aspects of eating disorders

As the eating disorders mainly affect adolescents and young adults, it is not surprising that their families have been the subject of special attention. This dates back to the earliest clinical descriptions of anorexia nervosa. For example, Lasegue in a classic account in 1873 wrote: '. . . it must not cause surprise to find me thus always placing in parallel the morbid condition of the subject and the preoccupation of those surrounding her. These are intimately connected and we should acquire an erroneous idea of the disease by confining ourselves to examining the patient'. The next year, William Gull, also commented on the family in relationship to treatment: '. . . the patients should be fed at regular intervals, and surrounded by persons who would have moral control over them; relations and friends being generally the worst attendants'.

Older accounts, mainly by physicians, usually saw the family as an intrusion on treatment and recommended that the patient be treated away from the sabotaging influence of the parents. During the 1950s and 1960s, psychoanalytic views dominated ideas about the role of the family in anorexia nervosa. Mother-daughter relationships were a major focus, while the father was usually regarded as distant and absorbed in his work. Notions of a psychopathogenic mother were explored, and individual psychotherapy recommended for both patient and mother to deal with their respective psychopathologies.

Modern accounts of the disorder give weight to family factors within a multifactorial aetiological model. Biological, psychological, and socio-cultural influences are acknowledged, and interact within the domain of family relationships. The disorder is seen within the context of the whole family system, embedded in its interactions. Models proposing linear causal relationships have been replaced by those emphasizing circular or recursive relationships in which the behaviour of each family member affects, and is affected by, the others. Several influential ideas concerning the family's role in the eating disorders are current, while a large number of research studies have been carried out since the 1980s to test them.

Clarification of the terminology regarding the subtypes of eating disorders is required since there may be differences in some of their familial associations. *Anorexia nervosa* (AN) refers to a disorder in which there has been substantial weight loss, a psychopathology characterized by a morbid fear of fatness, and a pursuit of thinness, and amenorrhoea. *Bulimia nervosa* (BN) is characterized by eating binges, attempted compensation for these by self-starvation, self-induced vomiting, or laxative abuse, and a morbid fear of fatness. If patients with anorexia nervosa also suffer from bulimia, we shall describe them as 'bulimic-anorexics'. Otherwise BN refers to patients who are around normal weight. Sometimes, patients with binge-eating who may not necessarily meet the criteria for BN have been studied; the condition here will be termed simply, *bulimia*.

GENETIC FACTORS

There is a familial aggregation of AN: around 6 per cent of patients have a sister who suffers or has suffered from the disorder. A large study comparing the incidence of eating disorders in relatives of patients with anorexia nervosa with those of a normal population sample has shown an increase in first-degree female relatives; AN: 4 per cent versus 0 per cent; bulimia 2.6 per cent versus 1.1 per cent (Strober 1991). The evidence for BN is less impressive, but Kassett *et al.* (1989) found a 10 per cent incidence of bulimia in female first degree relatives of patients with BN compared with 3.5 per cent in the relatives of a control group. A reasonably large twin study revealed a concordance rate of 56 per cent in monozygotic (MZ) twins compared with 5 per cent in dizygotic (DZ) twins (Holland *et al.* 1988). There are more limited data which suggest that the liability to inherit BN is less than AN with concordance rates for BN generally not differing between MZ and DZ twin paris (35 per cent versus 29 per cent), and quantitative genetic analysis of their families indicating that inherited factors are not significant. However, a recent twin study based on the large Virginia Twin Registry in the US found a probandwise concordance of 22 per cent in MZ pairs compared with 9 per cent in DZ pairs (Kendler *et al.* 1991). A genetic model provided a good fit to the results but a familial/environmental alternative was not much inferior.

The relationship between eating and affective disorders has also been of interest, particularly since some have argued that the former may be a 'variant' of the latter. A two to three fold increase in the life-time risk for affective disorders has been found in the relatives of patients with AN or BN compared with the general population. Some studies have suggested that this increase may be limited to patients with an eating disorder who also show evidence of depression. Eating disorders are not more common in relatives of patients with affective disorders. A single common transmitted liability seems unlikely.

It is not known what comprises the transmissible influence for AN. It may still be environmental rather than genetic. Those favouring a genetic factor have suggested, based on what is known about the genetics of personality, that the inherited phenotype may involve personality characteristics, perhaps avoidant (harm avoidance, low novelty-seeking) or compulsive traits (rigid, perseverative behaviours). A youngster with such traits may be especially vulnerable around puberty when uncontrollable physical and psychological changes occur, and may seek to impose even greater controls on his or her behaviour (Strober 1991). An alternative suggestion, that the phenotype involves a delay in psychosexual development, has been proposed because a comparison between discordant MZ and DZ twin pairs indicates that MZ pairs are similarly delayed and significantly different to the DZ pairs. Differences in neurotic traits were not found (Treasure and Holland, in press).

Caution must be exercised in interpreting the data because of small numbers and the possibility of an ascertainment bias where clinic-based twin samples commonly show a higher concordance rate than community based samples.

SOCIOCULTURAL FACTORS

AN is a disorder primarily of westernized societies. Recent studies of eating disorders in non-Western cultures suggest that a conflict between traditional and western values, often played out between the generations in the family, may be associated with an increased risk, especially of bulimia. This has been supported by studies comparing Greek girls in Greece and Germany, and in Asian girls in Pakistan and Bradford, England. In the last study the prevalence of bulimia in the Asian girls in Bradford was higher than in the native-born population (Mumford and Whitehouse 1988). There is also limited evidence, based on small groups of Afro-Carribean or Asian patients with BN in English clinics, that they are more likely to have experienced emotional deprivation and to have come from broken homes than native-born patients.

It is commonly believed that AN is more common in higher social classes. Studies of patients treated at specialist eating disorder centres supported this, but when patients are ascertained through catchment based case-registers where selection factors are less, the social class bias tends to disappear.

PSYCHIATRIC AND PHYSICAL DISORDERS IN OTHER FAMILY MEMBERS

An increased incidence of eating and affective disorders has already been mentioned. There is no over-representation of schizophrenia in families of patients with eating disorders. It has been suggested that substance abuse is more common; in some series, 12–19 per cent of fathers and about 5 per cent of mothers have been assessed as having a drinking problem. A higher rate of alcoholism than in controls has recently been confirmed in a study of first-degree relatives of 62 women diagnosed as having AN 10 years previously (Halmi *et al.* 1991). The incidence may be higher in the parents of bulimic compared with those of anorexic patients.

Kalucy *et al.* (1977) found that 20 per cent of both mothers and fathers in a series of 56 patients showed evidence of 'weight-related pathology'. Anecdotal reports often mention weight and eating preoccupations in these families, including odd diets, an emphasis on beauty and physical fitness, and a reliance on appearance as a measure of self-worth. Kalucy *et al.* also found a high prevalence of psychosomatic disorders in their families, for example, migraine in 30 per cent of mothers. However, in evaluating such accounts note must be taken of selection factors leading to particular patients being referred to specialist clinics. Anne Hall in New Zealand surveyed a normal population of 204 mothers of girls in the age range of those presenting with eating disorders. The mothers too had experienced a considerable number of illnesses including a high incidence of migraine and asthma. Thirty five per cent of the girls had obese or underweight relatives. A high proportion of mothers had experienced considerable weight fluctuations over their lives. It is thus possible that such features are no more common in families of eating disorders than in the general population. A consistent finding has been that patients with BN are more likely to have overweight parents, especially mothers, than patients with AN.

The aforementioned study of Halmi *et al.* also found a higher life-time incidence of obsessive-compulsive disorder and psychosexual problems (inhibited sexual desire or excitement) in the mothers of patients compared to controls. Fathers, however, showed no greater incidence of psychiatric disorders than control fathers.

The small number of studies of parents of patients with eating disorders have usually not found an increase of psychological disturbance when measured using standard instruments such as the Eysenck Personality Inventory.

FAMILY CHARACTERISTICS

Demography

Sociocultural factors have already been mentioned. Religious affiliation, family size or composition, and sibling position in the birth order generally do not differ when compared to the general population. Older parents at the time of the patient's birth seems a consistent finding both in AN and BN, but there has been no correction for a possible confounding effect of social class. The significance of this observation is unclear. It may perhaps reflect ambivalence about having children, or an older age at marriage related to difficulties in establishing psychosexual relationships, or greater difficulties in understanding or in achieving flexibility with an adolescent youngster. A few studies have indicated a preponderance of female siblings in families of patients with an eating disorder, but most have not.

Parental rearing

Clinical observations

The ideas of Bruch (1973) have been powerful. She stressed the development of a pervasive sense of ineffectiveness in AN, sufferers feeling powerless to control their bodies, or to be self-directed or autonomous. Also described was an inability to discriminate between bodily sensations, including hunger and satiety, and to differentiate between bodily tensions and psychological states such as anxiety or depression. Bruch regarded these impairments as having arisen out of early interactions between patient and mother. Recognition of bodily sensations such as hunger is learnt through this interaction, and depends on appropriate responses by the mother to the child's needs. If these have been disregarded or inappropriately dealt with by the mother, the child may become perplexed about her bodily sensations and their differentiation from emotional states. Eventually she is unable to identify these inner states. Bruch claimed that while the mothers of AN patients provided excellent physical care, it had been imposed in accordance with idiosyncratic concepts rather than in reaction to cues emanating from the child. An integrated self which knows what it feels and wants is thus not achieved, and difficulties become apparent at adolescence when demands for independence commence. Many clinical accounts support these observations concerning the patient's struggle to exercise control, but there have been few inquiries aimed at reconstructing early mother-child experiences.

Research

Several studies have applied the Parental Bonding Instrument, a self-rating questionnaire which seeks information about the respondent's recollections of both mother and father during the first 16 years of life. Two scales are derived, 'care' (reflecting warmth, affection, and empathy) and 'protection' (reflecting parental control, intrusion, and overprotection). Neurotic patients consistently show different responses, high 'protection' and low 'care', compared with healthy subjects. Results in patients with AN or BN have been inconsistent. Lower scores on 'care' for mothers or fathers or both, compared with controls, have been noted, but the variation in scores within the patient groups is notable (Palmer *et al.* 1988). Differences on 'protection' have been less evident. One study found entirely normal scores for a group of anorexic adolescents, while a similar aged group referred with other psychiatric disorders were low on 'care' and high on 'protection'.

A recent comparison by Schmidt and colleagues at the Institute of Psychiatry in London of the childhood experiences of large numbers of AN patients with bulimic patients, determined by a standardized semistructured interview, has pointed to a greater degree of disturbance in the latter, including more parental indifference, intrafamilial discord, physical abuse, and violence to other family members. The incidence of sexual abuse did not differ. Controls were not included but the rate of adverse childhood experiences was higher in the patients than reported for the normal population.

An issue related to childhood rearing is sexual abuse. About 30–40 per cent of patients with eating disorders report episodes of sexual abuse as children, but this is in the range reported by other psychiatric patients, and even normal populations (Pope and Hudson 1992). In one study of a sample from a college in the USA, a weak association was found between reported child sexual abuse and scores on the Eating Disorders Inventory. The relationship between sexual abuse and psychopathology is complex. For example, such abuse may be more likely in families also disturbed in other ways; or the family's way of dealing with the episode may be important in ameliorating or exacerbating the sequelae. In one study there was evidence that abnormal eating attitudes were more marked in subjects of abuse who also perceived their parents as less reliable.

Family functioning

Clinical observations

A systems view of the eating disorders in relation to family functioning has achieved dominance. Accounts by Minuchin and his team, Selvini-Palazzoli and colleagues, Stierlin, White, and the Maudsley group overlap, but with differences in emphasis. Observation of families of patients with AN have predominated; less attention has been devoted to those with sufferers from bulimia.

'Structural' perspective The views of Minuchin and his colleagues in Philadelphia (Minuchin *et al.* 1975, 1978) encompass the functioning of the whole family within the framework of an open systems model. This holds that certain types of family organization are closely related to the development and maintenance of symptoms, and that the symptoms at the same time play a major role in maintaining family 'homeostasis', that is in

regulating family interactions so as to preserve a family balance. The model is non-linear in terms of causation, instead involving circular or recursive processes based on feedback loops. Also postulated is a vulnerability on the part of the patient which may be physiological, or derive from a significant family preoccupation with food-related matters. The specifics of symptom choice are not emphasized in the model; its perpetuation is the chief focus. A precipitating event threatening the family's equilibrium is usually identified. This may be external or, more commonly internal, a family life-cycle transition requiring a change in family relationships.

The Minuchin group described the characteristics of what they termed 'the psychosomatic family', an unfortunate appellation since 'psychosomatic' is such an ill-defined notion. Their observations concerned the families of youngsters with AN, unstable diabetes, and asthma. Four characteristics were identified which together 'describe a general type of family process that encourages somatization': (i) *Enmeshment*: overclose relationships in the family typically manifested by frequent intrusions on each others' thoughts and feelings, speaking for each other, and 'mind-reading'. Closeness and loyalty are valued more than autonomy and self-realization. Subsystem boundaries are weak and easily crossed. The parental executive dyad may be powerless. Children may join one parent in criticizing the other, or may take inappropriate parental roles. One parent may enlist the help of a child in a battle against the other parent. (ii) *Overprotectiveness*: mutual concern becomes overprotectiveness; the parents 'guard' the children and the children become 'parent-watchers'. Nurturing and caring responses are constantly elicited. Signs of distress signal the approach of a possible conflict and call for comforting or distraction from other family members. When the youngster attempts to assert herself, she become the subject of a 'benign' form of concerned control. (iii) *Rigidity*: a limited and fixed pattern of interaction prevails which does not change when circumstances (either external or related to the family life cycle) require it. There is a commitment to maintaining the *status quo*. Issues that threaten change, such as negotiation of individual autonomy, are not allowed to surface. There is an attempt to absorb change rather than to adapt to it. (iv) *Lack of conflict resolution*: despite the submerged tension arising from the preceding characteristics of the family, overt conflict is avoided. If it does emerge it is quickly pushed aside and thus never resolved. Conflict is detoured, often involving an aspect of the youngster's illness. For example, there may be an emotional outburst by the patient when conflict threatens which thus distracts the family by eliciting shared concern. The involvement of the patient in the marital relationship is an important focus.

'Systemic' perspective   This group of descriptions flows from the work of

Gregory Bateson and his colleagues of the Palo Alto group, Palazzoli (1978), and Stierlin and Weber (1989). They are difficult to summarize since they are less contained and specific than the 'structural' model. Many features of the latter are endorsed, for example, enmeshment and lack of conflict resolution, but 'systemic' models tend to emphasize other aspects. These include family belief systems (creeds, 'myths') which limit the range of choices available to family members and prescribe particular roles for them; transgenerational patterns and their transmission; the 'rules' governing behaviour in the family; how circular interactions in the family's behaviour maintain the symptoms and the way in which the symptoms regulate family transactions (the 'function' they serve); and the processes underlying change and resistances to it. Systemic theorists are less concerned than their structural counterparts with family organization, hierarchies, and subsystem boundaries.

As an example of the systemic approach, consider the observations of Stierlin and Weber (1989). They emphasize a tight 'binding' within the family through powerful, rigid, and often implicit family beliefs, sometimes transmitted across several generations. Common beliefs are 'giving is better than receiving', 'I only feel OK if others feel OK', or 'you must not let yourself go'. The family is thus dedicated to loyalty, closeness, fairness, subjugation of individual needs, self-sacrifice, control of impulses, and avoidance of expressions of 'bad' feelings. Losses in previous generations may further foster this sense of family loyalty. Parents may be linked to their own parents through self-denial and self-sacrifice. Intense binding blurs boundaries between individuals in the family – 'all the rooms are interconnected and all the doors are open' – while influences from the outside world are kept at bay. Open alliances between any two members are prohibited because this represents disloyalty to the others. This may extend to a poorly defined parental hierarchy, resulting in a power vacuum which is contested through underground battles fought according to family rules promulgating self-denial or martyrdom. Individuation of an adolescent is difficult because the expression of individual needs breaks the family creed of self-denial, as does the prospect of separation and leaving home. The loss of grandparents may make the youngster even more guilty about abandoning her parents as she wonders whether they will survive her separation from them.

With the development of AN, the patient is on the one hand a strong contender in the 'self-denial stakes', but is also able to seek revenge for the suppression of her needs or the sense of betrayal by a parent or parents ('I need something but you can't give it to me'). The disorder may also save the family from the emergence of potentially uncontrollable emotions, for example those related to sexuality, while it also unites other members in a common family cause.

Maudsley perspective   This group has been influenced by both structuralist and systemic descriptions, and also by research studies (discussed below) of patients and families treated in a specialist eating disorder unit (Dare and Szmukler 1991). Consequently, their approach is more eclectic, with less inclination to describe family stereotypes, and more stress on particular problems in particular families. Instead of one of 'pathological functioning', a developmental perspective predominates. The family responds to the difficulties associated with its member's eating disorder by reverting to patterns which typified, and were indeed appropriate to, an earlier phase of family development. They somehow become stuck in this position. Balancing the needs for stability and change may be complicated by transgenerational rigidities and an inability to find new ways of acknowledging closeness and loyalty in the next phase of development. Research findings suggest that family members feel insufficiently close and flexible. Enmeshment and conflict avoidance may be responses to this. The important physical and psychological consequences of the patient's emaciation (for example impaired attention, concentration, memory) are recognized as constraining her autonomy. The family's fear for the patient's life and bewilderment in the face of her severe emaciation are respected.

There have been few clinical descriptions of family functioning in BN. Many features mentioned above are again noted (Schwartz *et al.* 1985) but heightened concerns with achieving success and attractiveness, as defined by the culture in which the family exists, have received special comment. However, a common impression prevails that the families of bulimics show more conflict, hostility, and disorganization.

Empirical studies of family functioning
To what extent have these plausible, clinically derived accounts of family functioning found support in research?

A substantial number of studies employing standardized measures of family functioning, including the FES, FAD, and FAM (see Chapter 2) have been conducted. However, difficulties arise in their interpretation. Patients have been drawn from varying clinical settings, involving selection factors that reflect referral patterns which are impossible to assess. They have been of varying ages and at different phases of illness which has, in addition, varied in severity. There may be differences between families having patients with AN or bulimia. Comparisons across different measures are hard to make because they may tap different dimensions of family life (see Chapter 2). Moreover, the relationship between these dimensions and those regarded as clinically meaningful is often unclear. Self-report measures are always prone to bias, especially in a clinical setting, due to denial or to a wish to present a normal family to the outside world. Where multiple family members have participated, a further problem arises in knowing what to make of discrepant perceptions. Most studies have employed a control group of normal young women, but their numbers have been relatively small; furthermore, the extent to which they have been matched on important factors affecting family functioning, such as social class or family composition, is often open to question. In all studies, family functioning at the time of investigation may reflect both pre-illness patterns of interaction as well as reactions to an ill family member. Aetiological inferences cannot be drawn with confidence. Finally, the link between family distress and the eating disorder tends to remain obscure. Even if families of eating disorder patients show greater levels of disturbance on certain dimensions, how specific are these for an eating disorder rather than for another psychiatric disorder.

Bearing these caveats in mind, the following seem to be the major findings. Nearly all studies have found that families with an eating disorder patient are more distressed or dysfunctional than control families. In most, but not all, studies which have examined the question, this has been more marked for families of patients with BN than AN. The major dimensions of poorer functioning have been in terms of what the standardized instruments measure: less cohesion, more conflict, more disorganization, less support, less expression of feelings, less disclosure, and less trust. Consistent patterns across studies are difficult to discern, and there is little evidence of specificity between family functioning and eating disorders. In some work families of patients with AN have shown features reminiscent of those described by Minuchin and colleagues, but others have found a heterogeneity of family 'types', or no differences on the relevant dimensions when compared with control families. (See Yager (1982), Kog *et al.* (1985), and Strober and Humphrey (1987) for a review of most of these studies.)

A few family *observational* studies have been conducted. Humphrey *et al.* (1988) compared families of bulimic-anorexics with normal controls using videotaped interactions rated according to Benjamin's Structural Analysis of Social Behaviour (SASB). The task was a discussion of a situation dealing with the daughter's separation from the family. Families of patients were more negative (ignoring, putting down, and 'walling-off'). They were also less helpful and trusting towards each other, and communication about the subject of autonomy was often contradictory. The results generally supported those from the self-report studies mentioned above. Humphrey (1989) conducted further research using the same methods but this time comparing subtypes of eating disorders. Families of restricting anorexic patients differed significantly from those with bulimia: parents in the former were more nurturing and comforting but also less attentive to the patient's statments, while in the latter there was more evidence of hostile enmeshment. Both groups also used ambiguous or contradictory communications more often than controls. The author concluded that the results were consistent with the theories of Bruch and Minuchin so far as AN was concerned, but differed when bulimia was present, hostile enmeshment predominating.

Three studies warrant closer consideration because of their attention to the selection of controls or special aspects of the methods used.

Dolan *et al.* (1990) compared the families of 50 bulimic women with 40 non-eating disordered women from the same community. Results were based on self-report using a Perception of Parents questionnaire and the authors' own scale measuring six family factors. There were no differences in family size, composition, or gender ratios. Parents of bulimic patients were significantly older at the time of the patient's birth. There was no difference in the parental divorce rate between the groups, but the patients reported less attention and affection from their parents, more parental marital disharmony, and more repressive sexual attitudes. There was a trend for more feminine stereotyping, but no differences on closeness between the respondent and other family members. The interesting finding that 22 per cent of patients' parents were of different ethnic origins, compared with 8 per cent of the controls, suggested to the authors communication problems in the family with emotional demands being expressed in non-verbal ways, for example, through food.

Rastam and Gillberg (1991) studied 52 youngsters around 16 years of age detected as having AN during the course of a population-based study in schools covering the city of Goteborg in Sweden. Each case was matched with a control of the same sex and age from the same school. Mothers of both groups were also interviewed. Parental age at the time of the child's birth did not differ in the two groups. Major family problems were significantly more common in the AN group (70 per cent versus 40 per cent). These included the death of a first degree relative, severe physical or mental disorders in close family, major parental secrets, parental disagreements, overly strict discipline, and poor problem-solving. A modified FACES questionnaire (see Chapter 2) was completed by the mothers. Differences on the major scales were small; however, statistically significant differences occurred on 10/24 items. These pointed to less family happiness and understanding, and less competence in problem-solving in the AN families.

The investigators also rated the characteristics of the so-called 'psychosomatic family'. There was a non-significant trend for enmeshment, rigidity, and conflict-avoidance to be more common in the AN families. Overprotection was more common in the controls. The authors concluded that fewer than one in five of the AN families had the 'psychosomatic family' constellation, and that no specific family dysfunction related to the eating disorder. This is an important study because of its careful selection of cases and controls, and because the cases were young and of recent onset (thus avoiding confounding by factors associated with chronicity rather than AN itself). A weakness was the reliance for many measures on the researchers' non-blind ratings.

In a complex set of investigations Kog and colleagues in Belgium also sought to examine the validity of the 'psychosomatic family' model in AN. An attempt was made to operationalize the four key characteristics (Kog and Vandereycken 1985; Kog *et al.* 1987; Kog and Vandereycken 1989). This involved extensive redefinition; for example, 'enmeshment' focused on the degree of variation in subsystems boundaries in carrying out different tasks. Fifty-five families were studied involving three tasks. A questionnaire attempting to tap the same dimensions was also administered. It was found that the 'insider view' (questionnaire) differed substantially from the 'outsider view' (observer's ratings). Cluster analysis produced seven family types, some similar to the Minuchin model, others not at all so. These investigators went on to examine another group of eating disorder patients and their families, comparing them with controls matched for social class, family size, and age and sex of the patient. The scales were further revised and renamed – now 'cohesion', 'adaptability', and 'conflict'. Again measurements were both observational and self-report. Compared with the controls, the eating disorder families showed significantly fewer disagreements between children and parents (taken as representing conflict avoidance), and less adaptability (taken as representing rigidity). However, there were differences *within* the eating disorder group, with AN families showing more 'cohesion' and less disorganization than those with bulimia. These studies, while carefully performed, are difficult to interpret. The derived scales are sometimes confusing and relationships to clinical notions of enmeshment, conflict avoidance and so on, appear tenuous. Overall, they provide limited support for the 'psychosomatic family'.

Expressed emotion

Derived from work in schizophrenia (see Chapter 6), EE has received attention in AN as well (Szmukler *et al.* 1985; Le Grange *et al.* 1992; van Furth *unpublished*) Parental critical comments are fewer in anorexic families than in those with a schizophrenic subject, possibly supporting the conflict-avoiding character of these families. However, emotional over-involvement is also low and not consistent with a high level of over-protectiveness. Families scoring high on EE may see themselves as less close than they would like to be, and to perceive their families as less adaptable. High EE is more common with BN. Parental criticism tends to be greater with a longer duration of illness and poorer social adjustment of the patient. Criticism, rated conventionally by interview with parents singly, is highly correlated with criticism expressed during a videotaped family meal.

Family factors and the course of the illness

In nine long-term outcome studies where the question has been examined, seven found a poorer outcome for AN when obvious family difficulties, especially poor relationships between the patient and her parents, have been evident. EE has also been shown to influence the prognosis. High parental EE has been shown to be associated with dropping out of family, but not individual, therapy, and with a poorer outcome for the illness at 6 or 12–18 months follow-up. In one prospective study where EE was measured before starting family treatment and again six months later, criticism increased in the poor outcome group but decreased in the good outcome group. The former also showed a significant decrease in warmth (Le Grange *et al.* 1992).

Marriage and patients with eating disorders

The proportion of patients with AN who are married may be increasing. As many as 20 per cent of patients at specialist clinics are married. Their marriages have been little studied. A retrospective survey has suggested that anorexics who marry have a later age of illness onset than those who remain single, and come from homes in which divorce and remarriage are more common (Heavey *et al.* 1989). Their families of origin tend to have less enmeshment and less conflict avoidance. Husbands of patients may show more than expected neurotic symptoms, but whether this is due to assortative mating or a reaction to their wives' disorder is unknown.

Clinical accounts also suggest that both marital partners remain closely bound with their families of origin and this may interfere with the establishment of marital intimacy. Both partners are likely to be dissatisfied with the marriage, husbands to a greater degree than their wives, perhaps reflecting the latter's propensity for denial. A distinction between couples in which the eating disorder is present at the time of marriage and those in which it occurs after may be useful. In the former, it is more likely that the eating disorder serves to stabilize the relationship. For example the patient may feel she deserves no-one better, while the husband may wish to care for someone with a problem, perhaps harbouring a fantasy of rescue. Where the eating disorder occurs after marriage, a crisis affecting family relationships is often a precipitant.

The incidence of premature births, low birth weight, and perinatal deaths are raised in pregnancies to patients who have not fully recovered nutritionally. The limited evidence to date suggests that the pregnancy and post-partum period are reasonably coped with and that infants gain weight normally. However, those who still have anorexic symptoms may find weight gain during pregnancy difficult. Subjects with BN usually improve during pregnancy because of fears of harming the fetus, but in most cases symptoms recur after delivery.

FAMILY THERAPY

Approaches to treatment ostensibly follow from the formulations described earlier. Many case reports have claimed successful outcomes with structural or systemic family therapy. Structural approaches involve manoeuvres challenging: (a) enmeshment (for example, discouraging speaking for others or 'mind-reading', strengthening subsystem boundaries, supporting parental authority, blocking parent – child coalitions); (b) rigidity (for example, disrupting repetitive interactional patterns); (c) overprotection (for example, blocking inappropriate 'rescues' and supporting coping behaviour); and (d) conflict avoidance (for example, blocking attempts to suppress conflictual discussions while encouraging problem-solving strategies).

Systemic therapies are more difficult to describe but usually involve maintaining strict therapist neutrality, circular questioning aimed at generating maximal information about and for the family, positive connotation of the family's behaviours, reframing their meaning (usually in the light of transgenerational issues or family 'myths'), and offering prescriptions for the family based on reframing which aim to make it difficult for the family to continue to behave as previously. Prescriptions may involve the paradoxical injunction not to change, acknowledging the risks of altering the existing equilibrium.

Minuchin *et al.* (1975) and Martin (1985) reported excellent outcomes for substantial groups of patients treated with structural family therapy, the vast majority having recovered on one to three years follow-up. However, both studies were uncontrolled, and involved young patients, most under 16 years of age, for whom the prognosis is known to be good in any event.

In a carefully controlled study at the Maudsley Hospital, Russell *et al.* (1987) randomized 80 patients to family therapy or routine individual supportive therapy on discharge from an inpatient weight-restoration program. Patients were stratified according to age of onset, duration of illness, and presence of bulimia. Therapy was provided for a year. Patients with an early onset of illness (less than 18 years) and a short history (less than three years) had a significantly better outcome at one year follow-up on all measures with family therapy. There was a trend for patients with a later age of onset (19 years or older) to fare better with individual therapy. Patients with a long history and those with bulimia responded equally to both treatments. Family treatment involved an average of 10 sessions compared to 15 sessions for individual treatment. High EE was associated with dropping out of family, but not individual, treatment and a poorer outcome (Szmukler and Dare 1991).

Family therapy in this study involved special attention to engagement, providing information about the disorder and the effects of starvation, acknowledging the parents' anxieties about their child's dangerous condition, reducing blame, and in initially helping the parents to take control of their daughter's diet. While resulting in weight gain, the last strategy also provided a focus for re-establishing the parental dyad in an effective executive role. As the patient's physical condition improved, therapy became increasingly concerned with normal adolescent issues of autonomy and individuation, and how these might be negotiated within the family. A structural approach, less directive and more reflective than that described by others, generally predominated, with strategic or systemic measures being introduced when progess faltered. For further details of the treatment see Dare and Szmukler (1991). The style of therapy was acknowledged by the investigators to be more suitable for younger patients living with their parents than for older patients less subject to obvious parental control.

Dare, Eisler, and colleagues have continued to explore the role of family treatments. A pilot study involving young patients, for instance, has suggested that seeing parents alone with interventions along the lines described above may be as effective as seeing the entire family. Studies evaluating other styles of family therapy for older and chronic patients are currently in progress.

Vandereycken (1987) has provided a useful account of the principles underlying a family-oriented approach in the context of multimodal treatment, including in-patient treatment, and cautions against inflexible attitudes which may easily lead to the clinician working against, rather than with, the family.

IMPLICATIONS FOR THE CLINICIAN

The consensus view of clinicians in the field is that a family orientation is integral to an understanding of the eating disorders. While family distress is common, research attempts to verify common clinical observations and widely held theories based on them have met with limited success. The task, however, should be recognized as exceedingly demanding. There is good evidence that a substantial family treatment component is useful for young patients living with their families. The role of family treatment for the others, including those with BN, is not yet clearly defined. Single parent, or other 'anomalous' family constellations, present a further complication.

In the light of the information reviewed in this chapter, it is clear that family factors need to be considered in the assessment and management of the eating disorders. This needs to be exercised in the context of a multidimensional approach which also heeds biological, psychological, and sociocultural issues. The following suggestions for the clinician who lacks specialist training in family work may prove useful.

A family assessment along the lines suggested in Chapter 3 will generally be illuminating. It is probably not profitable to look for particular family 'types' since the evidence points to considerable heterogeneity in family relationships. An individual assessment which is open to the possibility that there is no significant dysfunction in the family is appropriate. The way in which family interactions have been moulded by the illness should be considered, and possible maintaining factors identified. Developmental and family life-cycle transitions often provide a relevant context. Experience suggests that a multigenerational history frequently leads to a better appreciation of the family's belief systems and traditions. These may not lead to aetiological hypotheses, but they are helpful in understanding the meaning of the illness for the family and why their responses to its demands have not met with more success. The integrity of interpersonal and subsystem boundaries should be considered, particularly that of the parental dyad. The patient's intrusion into the marital bond usually has implications for treatment.

As family factors are important in influencing treatment outcome, we suggest that the clinician involve the family in the treatment plans. The work of the Maudsley group points to some useful approaches applicable to the non-family therapy specialist. The family should be engaged in a positive manner, as allies in the task of helping their daughter. An opportunity to talk about the relatives' fears and attempts to help the patient encourages collaboration, as does the provision of information by the clinician about the illness and its consequences. Even if the family is not to be directly involved in treatment, an explanation of the plan and its rationale fosters their support. Not infrequently, the family is the most powerful agency in ensuring admission, if required, of a reluctant patient. The issue of blame should be addressed and the family told that families do not cause eating disorders.

Referral to a family therapist with experience in treating eating disorders is appropriate with patients for whom such treatment has been shown to be useful. With less severely ill youngsters, and where the family is well-functioning and supportive, the clinician may consider family oriented treatment himself. This may proceed along the lines described by the Maudsley group. Following engagement and explanation to the patient and her family of the importance of weight gain to restore the patient's capacity for autonomous functioning, the parents, acting together, may be encouraged to assume control of the patient's eating. Advice may be given about caloric requirements, perhaps with the help of an experienced dietician, and the parents supported in developing a strategy for insisting that their daughter eat what they provide. A united couple will often prevail over a resistant daughter. They will also often be successful in prohibiting exercise and other behaviours directed at losing weight. Success in achieving weight gain will strengthen parents' confidence in their parental role and the bond between them. The parental subsystem is defined and any patient intrusion into the marital relationship, countered. As weight increases, the focus will shift to negotiation between patient and parents about normal adolescent issues as previously described. In adopting this treatment the clinician must be able to join with the patient as well as her parents; he will make it clear that she must first be 'rescued' from her dangerous emaciation by her parents before he can help her to grapple with the more important battles, those to do with establishing her independence.

With older, chronic patients, or those with BN, the clinician may encourage a relationship between patient and parents in which the eating disorder plays as little role as possible, while pursuing individual treatment with the patient. Even with a predominantly individual approach, a knowledge of the family will often inform important aspects of the treatment. For example, if the patient is 'triangulated' in the parental dyad, interventions aimed at helping her to disengage from this may be explored, including among other possibilities, developing relationships outside the family, keeping secrets from the parent with whom she has a coalition, or testing the strength of the parental relationship by challenging behaviour. If the parents are seen, they may be encouraged to demonstrate to the patient that they have a private life together and can live harmoniously without her assistance.

We hope these suggestions will prove useful in providing a framework for family involvement in the treatment plan. Advice from a consulting family therapist may be necessary if uncertainties arise.

REFERENCES

Bruch, H. (1973). *Eating disorders: obesity, anorexia nervosa and the person within*. Basic Books, New York.

Dare, C. and Szmukler, G.I. (1991). Family therapy of early-onset, short-history anorexia nervosa. In *Family approaches in treatment of eating disorders* (ed. D. Blake Woodside and L. Shekter-Wolfson), pp. 23–48. American Psychiatric Press, Washington DC.

Dolan, B.M., Evans, C. and Lacey, J.H. (1989). Family composition and social class in bulimia. A catchment area study of a clinical and a comparison group. *Journal of Nervous and Mental Diseases*, **177**, 267–72.

Dolan, B.M., Lieberman, S., Evans, C. *et al.* (1990). Family features associated with normal body weight bulimia. *International Journal of Eating Disorders*, **9**, 639–47.

Halmi, K.A., Eckert, E., Marchi, P. *et al.* (1991). Comorbidity of psychiatric diagnosis in anorexia nervosa. *Archives of General Psychiatry*, **48**, 712–18.

Heavey, A., Parker, Y., Bhat, A.V. *et al.* (1989). Anorexia nervosa and marriage. *International Journal of Eating Disorders*, **8**, 275–84.

Holland, A.J., Sicotte, N., and Treasure, J. (1988). Anorexia nervosa: evidence for a genetic basis. *Journal of Psychosomatic Research*, **32**, 561–71.

Humphrey, L.L. (1988). Relationships within subtypes of anorexic, bulimic, and normal families. *Journal of the American Academy of Child and Adolescent Psychiatry*, **27**, 544–1.

Humphrey, L.L. (1989). Observed family interactions among subtypes of eating disorders using structural analysis of social behavior. *Journal of Consulting and Clinical Psychology*, **57**, 206–14.

Kalucy, R., Crisp, A., and Harding, B. (1977). A study of 56 families with anorexia nervosa. *British Journal of Medical Psychology*, **50**, 381–95.

Kassett, J.A., Gershon, E.S., Maxwell, M.E. *et al.* (1989). Psychiatric disorders in the first-degree relatives of probands with bulimia nervosa. *American Journal of Psychiatry*, **146**, 1468–71.

Kendler, K.S., Maclean, C., Neale, M. *et al.* (1991). The genetic epidemiology of bulimia nervosa. *American Journal of Psychiatry*, **148**, 1627–37.

Kog, E. and Vandereycken, W. (1985). Family characteristics of anorexia nervosa and bulimia. *Clinical Psychology Review*, **5**, 159–80.

Kog, E. and Vandereycken, W. (1989). Family interaction in eating disorder patients and normal controls. *International Journal of Eating Disorders*, **8**, 11–23.

Kog, E., Vertommen, H., and Vandereycken, W. (1987). Minuchin's psycho-somatic family model revised: A concept-validation study using a multitrait-multimethod approach. *Family Process*, **26**, 235–53.

Le Grange, D., Eisler, I., Dare, C., and Hodes, M. (1992). Family criticism and self-starvation: a study of expressed emotion. *Journal of Family Therapy*, **14**, 177–92.

Martin, F.E. (1985). The treatment and outcome of anorexia nervosa in adolescents: A prospective study and five year follow-up. *Journal of Psychiatric Research*, **19**, 509–14.

Minuchin, S., Baker, L., Rosman, B.L. *et al.* (1975). A conceptual model for psychosomatic illness in children. *Archives of General Psychiatry*, **32**, 1031–8.

Minuchin, S., Rosman, B.L., and Baker, L. (1978). *Psychosomatic families: anorexia nervosa in context.* Harvard University Press, Cambridge (Mass).

Mumford, D.B. and Whitehouse, A.M. (1988). Increased prevalence of bulimia nervosa among Asian schoolgirls. *British Medical Journal*, **297**, 718.

Palazzoli, S.M. (1978). *Self-starvation: from individual to family therapy in the treatment of anorexia nervosa.* Jason Aronson, New York.

Palmer, R.L., Oppenheimer, R., and Marshall, P.D. (1988). Eating-disordered patients remember their parents: A study using the Parental-Bonding Instrument. *International Journal of Eating Disorders*, **7**, 101–6.

Pope, H.G. and Hudson, J.I. (1992). Is childhood sexual abuse a risk factor for bulimia nervosa? *American Journal of Psychiatry*, **149**, 455–63.

Rastam, M. and Gillberg, C. (1991). The family background in anorexia nervosa:

A population-based study. *Journal of the American Academy of Child and Adolescent Psychiatry*, **30**, 283–9.

Russell, G.F., Szmukler, G.I., Dare, C., and Eisler, I. (1987). An evaluation of family therapy in anorexia nervosa and bulimia nervosa. *Archives of General Psychiatry*, **44**, 1047–56.

Schwartz, R.C., Barrett, M.J., and Saba, G. (1985). Family therapy for bulimia. In *Handbook of psychotherapy for anorexia nervosa and bulimia* (ed. D.M. Garner and P.E. Garfinkel), pp. 280–310. Guilford, New York.

Stierlin, H. and Weber, G. (1989). *Unlocking the family door: a systemic approach to the understanding and treatment of anorexia nervosa*. Brunner/Mazel, New York.

Strober, M. (1991). Family-genetic studies of eating disorders. *Journal of Clinical Psychiatry*, **52**, 9–12.

Strober, M. and Humphrey, L.L. (1987). Familial contributions to the etiology and course of anorexia nervosa and bulimia. *Journal of Consulting and Clinical Psychology*, **55**, 654–9.

Szmukler, G.I. and Dare, C. (1991). The Maudsley study of family therapy in anorexia nervosa and bulimia nervosa. In *Family approaches in treatment of eating disorders* (ed. D. Blake Woodside and L. Shekter-Wolfson). American Psychiatric Press, Washington D.C.

Szmukler, G.I., Eisler, I., Russell, G., and Dare, C. (1985). Anorexia nervosa, parental 'expressed emotion' and dropping out of treatment. *British Journal of Psychiatry*, **147**, 265–71.

Treasure, J. and Holland, A. (in press). Genetic models of eating disorders. In *Handbook of eating disorders: theory, practice, and research* (ed. G.I. Szmukler, C. Dare, and J. Treasure). John Wiley and Sons, Chichester.

Vandereycken, W. (1987). The constructive family approach to eating disorders: Critical remarks on the use of family therapy in anorexia nervosa and bulimia. *International Journal of Eating Disorders*, **6**, 455–67.

Yager, J. (1982). Family issues in the pathogenesis of anorexia nervosa. *Psychosomatic Medicine*, **44**, 43–60.

8

The family and physical illness

The relevance of the family to the physically ill patient is supported by the following claims (Turk *et al.* 1985):

1. A variety of health-promoting and illness-preventing behaviours (for example, smoking, alcohol consumption, and dietary habits) in children are modelled on similar patterns of behaviour in parents.

2. A person's definition of his illness (that is, as to its causes and consequences) is largely derived from consultation with family members.

3. Agreement among family members' conception of illness has been related to successful treatment outcome.

4. Family attitudes are a major factor in patient compliance with recommended treatment regimes.

5. Chronic illness in a family member adversely affects other family members as well.

6. Maladaptive family interaction patterns can affect the course of both acute and chronic illness (that is, problem maintenance). As will be discussed below, family interaction patterns have also been implicated as a causal or precipitating factor in illness.

7. Studies suggest that over 70 per cent of sickness episodes are handled outside the formal health care system, and self-treatment within the family provides a substantial proportion of health care.

Thus the family may be viewed as a resource for preventing illness, assisting appropriate coping, and enhancing compliance with treatment and rehabilitation. The family may also be viewed as a causal or precipitating factor or as an influence upon chronicity or relapse. The reciprocal connections between patient, family, and medical and nursing staff all require consideration according to a family systems view of illness. Furthermore, cultural and social factors in a patient's family of origin may strongly influence the way illness is expressed (for example, pain). Contextual factors (for example, the environment of a coronary care unit or changes to the household necessitated by caring for a quadriplegic person or by home dialysis) will also influence the family's response to the illness.

It has also been noted that within a given family particular members tend to fall ill much more often than others, and that some families appear to consume a disproportionate share of medical, psychiatric, and social services in a community.

Research considerations

Family oriented research in health care has to contend with multi-dimensional data about a large number of interrelated variables, usually derived from small sample sizes. The progress of the past decade towards resolving the many methodological and conceptual problems in this area of research is usefully reviewed by Patterson (1990). Two broad approaches, model-driven and empirical-driven dominate the field (Fisher *et al.* 1990); these are derived from systems theory and social epidemiology respectively. The strengths and weaknesses of each approach have been described by Campbell (1986).

Other difficulties include uncertainty about what constitutes normal or pathological responses to illness, and the direct impact of the treatment environment on a patient's thinking and emotions. Moreover some medical treatments and psychosocial interventions may benefit the patient but be detrimental to his family (for example, home dialysis) (Steinglass and Horan 1988).

The customary research approach has been to consider a particular illness from various perspectives, for example, case studies, family demography, life events, family interaction and family function (see Chapter 2), applied cross-sectionally and prospectively. The sheer number of diseases and their great variability over time and in different patients, have limited the utility of such approaches.

An ambitious attempt to bring order into the field has been proposed by Rolland (1984, 1987). According to his model different illnesses share common characteristics which produce a particular type of stressor for a family. These include:

1. Illness onset – an illness of acute onset (for example, myocardial infarction or stroke) requires prompt mobilization of personal resources, alteration of roles, and effective use of external supports. By contrast an illness of gradual onset, for example, emphysema or Parkinson's disease, allows more time for the family to adjust.

2. Course of illness – chronic illnesses test the family's capacity over an extended period. During this time different coping responses may arise because of the course of the illness. Thus a progressive course

with increasing disability or where remission from illness is virtually non-existent (for example, Alzheimer's disease, juvenile diabetes) may be handled differently than one which is constant and steady after an acute phase (for example, stroke or spinal cord injury), or one which is relapsing or episodic (for example, ulcerative colitis, multiple sclerosis). Each pattern makes different demands on the family to change or maintain particular roles and relationships.

3. Outcome – both predictability of illness outcome and its impact on life-span will also influence the family's response. Thus a diagnosis of metastatic carcinoma carries with it the likelihood of imminent death. By contrast idiopathic epilepsy or congenital blindness is compatible with a full life-span; juvenile diabetes lies between the two extremes.

4. The nature and degree of incapacity – the handicap may be in cognition (for example, Alzheimer's disease), sensation (for example, blindness), mobility (for example, stroke, multiple sclerosis), physical energy (for example, cardiovascular and respiratory disease), appearance (for example, burns, neurofibromatosis), social desirability (for example, colostomy). Certain illnesses confer multiple deficits (for example, stroke, AIDS).

The second dimension in Rolland's model is that of time-phases in the course of an illness, that is, episodes of crisis, a period of chronic stability, and the terminal phase.

Family function constitutes the third dimension. This includes an assessment of family structure (see Chapter 2), the life-cycle stage of the family, which involves its capacity to meet stage-appropriate tasks, for example, separation-individuation, marriage, child-rearing, etc; the family's belief system, particularly about the illness; transgenerational patterns of illness; practical matters such as financial, vocational, accommodation difficulties, and changes in daily routine for caring for other family members; alteration in leisure and recreation patterns, in sleeping arrangements, and in the structure of household furniture to accommodate the sick person; and contacts with friends and social agencies.

Thus biologically dissimilar illnesses may still share strong similarities in terms of the psychosocial impact they exert on the family's adaptive responses.

The meaning of the illness for the family

An ill person often seeks an explanation for his illness in terms of 'why'?, 'why me?', 'why now?'. Family members may do likewise. Liaison psychiatrists have described the meanings that people may give to illness, for example, illness as punishment, challenge, weakness, the enemy, relief, an interpersonal coping strategy, irreplaceable loss, or a source of new values (Lipowski 1970). The narrative quality of these views, their origins in personal, family, and tribal history and their role in medical care have been studied by medical anthropologists (Kleinman 1988) and family therapists (Wynne *et al.* 1992). Family researchers in an urban hospital setting for instance have noted that most patients with an acute illness accept a biomedical explanation for it but this is much less so with chronic illness (Jaber *et al.* 1991). The fears and uncertainty that patients hold about their illness are strongly influenced by their family's previous experience of illness, often extending back across generations, a part of the family's identity. An informative account of families' 'medical myths' which contribute to their difficulty in coping with the care of a sick child is provided by Hardwick (1989); and similar considerations apply to adult patients. Clinicians too, may have experienced illness in their own family which affects their capacity to help their patients (Mengel 1987).

In addition to belief systems about illness, family members may share assumptions about their social world in general (for example, to what degree is it trustworthy, predictable, or potentially dangerous?). These 'world views' and the family's perception of the abilities of its members and the family as a whole to interact with the world affect the coping responses of the family. These aspects appear to differentiate families which make effective use of family oriented hospital treatment programmes (Costell and Reiss 1982) and correlate with mortality rate in home-dialysis treatment for chronic renal disease (Reiss *et al.* 1986).

Illness may act as a chronic stressor on the family causing a recurrent periodic sequence of crisis (that is, disruption of patterns of interaction) followed by adaptation (that is, new learning or second-order change in response to the stressor), followed by adjustment (that is, stability) until the next crisis period. Crises may be due to deterioration in the patient's condition, financial and other practical difficulties, or changes in family structure and relationships. These patterns of stress lead to changes at levels of the individual, family, and wider social system. A model of this process with considerable research potential has been proposed (Patterson 1988).

Successful adaptation to illness is associated with two constructs of family function, role flexibility and family rules which permit emotional expressiveness (Koch 1985). The long-term efficacy of therapeutic family intervention based on these constructs requires evaluation, but extant early findings will be summarized below. Such intervention known as 'medical family therapy' aims to enhance the role of the patient and family members as informed, active participants in the treatment process and to reduce the morbidity due to the illness' impact on family life (McDaniel *et al*. 1993).

By way of example of these general considerations, we now examine the influence of family factors in an acute physical condition, myocardial infarction, and in several chronic diseases.

ACUTE PHYSICAL ILLNESS AND THE FAMILY – THE CASE OF MYOCARDIAL INFARCTION

Family influences in aetiology

The genetic influence in the aetiology of coronary artery disease, mediated through cholesterol metabolism, obesity, proneness to atherosclerosis and hypertension is beyond the scope of this chapter, but the preventative role of the family in respect of these risk factors will be described below.

Since the 1950s psychosocial research has concentrated on four areas relevant to coronary artery disease – life events, personality variables, family (especially spousal) interactions, and sociodemographic characteristics. The first three are particularly relevant to our purpose.

1. Life events – clinicians have long identified the temporal relationship between onset of illness and experience of loss and separation, especially when loss leads to 'helplessness – hopelessness' and 'giving-in, giving-up' responses (Engel 1968). Life-events research has attempted to examine this systematically leading to scales to measure events, for example, the Schedule of Recent Experiences (SRE) which assigned a stressor score to specific events in a person's life (Holmes and Rahe 1967). The three most stressful events (death of a spouse, divorce, and marital separation), as well as 13 other items in the list of the 33 most significant SRE events are directly connected with family relationships. Specific schedules for particular illnesses, for example, myocardial infarction, have been devised. It should be noted that not all events on these scales are undesirable, yet they are deemed to be stressful.

Numerous conceptual and methodological objections have been lodged against this kind of research (Campbell 1986; Miller *et al*. 1986). However, one consistent finding is the increase in both incidence of and mortality from coronary artery disease following the death of a spouse, an effect most evident among widowers in the 12 months following bereavement.

Three hypotheses have been proposed for this phenomenon (Campbell 1986):

(a) Homogamy, a person consciously or unconsciously selects a marital partner who shares similar characteristics, including personality and behavioural traits which may affect health and longevity. Hence both partners tend to die at a similar age.

(b) The couple shares an unfavourable physical environment, including exposure to dietary and other pathogenic factors.

(c) Mortality is mediated by neuro-endocrine and psycho -immune factors activated by the emotional response to the loss.

2. Personality vulnerability and family of origin – in the 1940s psycho-analysts reported on personality characteristics which were prominent in the 'typical' myocardial infarct patient. Such a person is extremely ambitious, has good organizing ability, and strives compulsively to achieve goals associated with power and prestige. He tries to control strong aggressive impulses but is often hostile in his social and professional life. His creativity is extremely limited and infantile aggression and oral dependence dominate his inner life. The unconscious dynamics of such a person were formulated in terms of an unresolved competitive relationship in childhood with a feared and envied parent. In adult life the person repeats the competitive pattern but he tries to master it by identifying with superiors and compulsively seeks self-validation from them.

Many elements of this account became incorporated in the description of the Type A personality, though most of the research into this subject describes cognitive, attitudinal, and behavioural characteristics, and neglects the developmental and family origins of Type A which were implicit in the earlier psychoanalytic account.

Of particular interest in the study of various dimensions of Type A behaviour is the construct of hostility and its possible association with an increased incidence of mortality from coronary artery disease. These findings have renewed interest in the family and social relationships of the Type A personality.

It appears that Type A individuals typically have a quantitatively large social network and score highly on scales of social interaction. However the quality of support received from these relationships as judged by the Type A subject is inversely correlated with Type A personality scores and the level of conflict in these relationships is perceived to be high. The relationships become sources of stress and are compounded by a fear of failure and doubts of self-worth; this further aggravates the person's hostility and Type A behaviour, thereby creating a spiral of interpersonal stress which may increase the risk of myocardial infarction (Price 1988).

3. Spousal characteristics – early family oriented studies examined the spouse's personality, usually in terms of the degree of concordance

between partners on Type A scores. Greater similarity was associated with a diminished risk of coronary artery disease among men. Conversely, Type A men married to Type B women (that is, who do not show Type A characteristics and have a more relaxed attitude to life) showed an increased risk of coronary artery disease, an association strengthened by a woman's educational attainment exceeding her husband's and by her employment outside the home. The wife's report of higher levels of work-related stress correlated with an increased risk of the husband developing symptoms (Eaker *et al.* 1983).

These associations have not been adequately explained. It has been suggested that the Type A husband has a more traditional view of gender relationships in a marriage, and thus is threatened by his wife's success. Or it may be that her work-related problems cause him extra anxiety, or that her involvement in the vocational and social demands of her work reduces her ability to act as a social buffer for her husband in his own Type A-induced stresses.

In a landmark study some 10 000 male civil servants in Israel aged 40 years and over, without clinical evidence of cardiac disease, were followed up for five years. A variety of family problems correlated with the development of angina, as well as other coronary risk factors such as an increased blood pressure, raised serum cholesterol, and abnormal ECG findings. No clear correlation between family problems and myocardial infarction was noted. Where a male subject rated highly on anxiety ratings, his perception of his wife's supportiveness was associated with a reduced risk of developing angina (Medalie and Goldbourt 1976).

Three possible interpretations may be made of the association between angina but not of myocardial infarction with family problems and spouse support in previously asymptomatic men (Campbell 1986):

(a) a type 2 error, that is, myocardial infarction is much less common than angina, so that the influence of family problems on myocardial infarction could not be detected;

(b) angina and myocardial infarction have different risk factors;

(c) angina is a symptom and therefore a form of illness behaviour. Anxious men who do not regard their wife as supportive or who perceive a great deal of family problems, may report more angina-like symptoms without necessarily having coronary artery disease.

Prospective studies which include men with an established history of coronary artery disease note that high levels of spouse and family support are correlated with a lower prevalence of heart disease at the beginning of the study, but do not influence the subsequent development (that is, incidence) of disease to a significant degree. A review of such studies concludes that the discrepancy between prevalence and incidence studies of coronary heart disease which look for a correlation with spouse supports or family problems may be due to the fact that the development of clinical features of heart disease in a person actually causes difficulties in family and social relationships (Campbell 1986).

4. Family interaction studies – numerous prospective studies demonstrate the protective effect that harmonious family relationships and strong social supports have on physical health (especially among males), and in reducing the mortality and morbidity rates of a variety of diseases, including coronary artery disease. Specific patterns of family interaction which may be pathogenic for coronary heart disease have not been identified. In essential (primary) hypertension, which is a significant risk factor for heart disease, family interaction studies suggest a link between hypertension in adults, obesity in children, and family patterns of emotional expressiveness and closeness of relationships. The association does not indicate the direction of causality; it merely suggests that family members' perceptions of their relationships are associated with psychophysiological responses and behaviour patterns which are linked to risk factors for coronary artery disease (Hanson *et al.* 1991).

Precipitation of myocardial infarction

In a study of 117 patients with no known history of pre-existing cardiovascular disease who developed life-threatening ventricular arrythmias, 25 had experienced acute emotional distress as an apparent precipitant (Reich *et al.* 1981). In people with a history of coronary artery disease emotional distress, including family conflicts, have been implicated in precipitating silent myocardial infarction (Rozanski *et al.* 1988).

A psychophysiological state, termed 'vital exhaustion' has been described as a precursor of myocardial infarction (Appels 1990). It is characterized by a subjective sense of loss of energy, social withdrawal, marked irritability, and strong feelings of annoyance. Retrospectively, patients report having felt a wish to be dead, but this state differs from the suicidal despair of depression. Furthermore, unlike depression, feelings of guilt and low self-regard are not evident. The feeling of personal exhaustion is associated with a number of chronic interpersonal difficulties, including serious conflict in the marriage or in the workplace, persistent concerns about children (in regard to their behaviour and school performance), and long periods of overtime at work. It has been suggested that the Type A individual who declines into a prolonged state of 'vital exhaustion' and alienates his family or social supports may be at particular risk of developing a myocardial infarction.

Impact of the family on the sequalae of myocardial infarction

The mortality rate following myocardial infarction is significantly greater in single than in married people. Distinctions among the various types of single lifestyle (that is, unmarried, widowed, or divorced) are often lacking in the research literature but the protective effect of marriage is clear. The difference in mortality applies both to the hospitalization period following myocardial infarction and to follow-up periods as long as 10 years post-infarct.

The mechanism by which this protective effect is mediated is uncertain. Studies in the 1960s indicated that men tended to deny the medical significance of ischaemic chest pain and delayed seeking appropriate treatment. A spouse may be able to challenge this denial, though it also appears that impersonal authority (for example, a policeman, foreman, or nurse at the workplace) may be more persuasive. This is consistent with characteristics of the Type A personality (Hackett and Cassem 1969).

A second possibility is that marriage may confer a survival advantage by its influence on lifestyle risk factors, for example, smoking, physical activity, and diet, and may exert their effects both by reducing the severity of the infarct, and by enhancing the patient's cooperation with rehabilitation especially when he persistently denies the significance of his condition (Bar-On and Dreman 1987) (see below).

A third possibility is that social isolation and loneliness may create social and emotional stresses which increases the mortality rate.

Impact of acute illness on the family

While much clinical and research attention has been given to the anxiety and distress an acute life-threatening episode of coronary ischaemia causes for family members, it should be remembered that not all interpersonal difficulties are the consequence of the infarct itself. Marital and family tensions may precede the acute illness, and these may continue or be exacerbated by the myocardial infarct, its practical consequences, and symbolic meanings. In a study of 113 patients with myocardial infarction admitted to a Coronary Care Unit, 40 were rated as experiencing significant emotional difficulties. Of these, 11 rated themselves as having experienced a similar degree of distress prior to the infarct episode, six described an exacerbation of pre-existing difficulties after the infarct, and 22 developed these problems following the infarct. Marital and relationship problems, sexual difficulties, and work-related stresses were the main areas of concern (Mayou 1989). This study also found that neither the clinical severity of the infarct nor demographic characteristics correlated with an adaptive psychological response by the patient.

Many effects of an acute illness on the patient are experienced in varying degrees by other family members. From a psychodynamically informed perspective (Strain 1978) these include threats to self-esteem and to the sense of intactness, separation anxiety, fears of loss of love and approval, concern about loss of control, anxiety about injury to body parts, guilt, and shame.

The response of family members to the patient in the face of these anxieties tests their ability to:

(1) accept the fact that the patient may regress physically and mentally;

(2) help the patient ward off the stresses evoked by the illness;

(3) tolerate the patient's expressions of his fears and feelings;

(4) enlist the patient's sense of basic trust, that is, his confidence that he will not be abandoned and will continue to be supported in his efforts to regain his health;

(5) mobilize outside resources as required by the patient.

Both the family's needs and resources influence its coping responses to acute illness. While the following summary of the needs of family members may seem like little more than commonsense medical practice it is easy to overlook them in the busy, alienating, high-tech environment of a modern hospital. Family members needs include (Power 1991):

1. To be listened to and seen as people who are attempting to cope with their anxieties, and so need to be given the opportunity to express ambivalence, guilt, and other feelings.

2. To be given information about the illness, its likely consequences and the treatment in ways that can be understood. Often family members may fail to understand the doctor's explanations because they are preoccupied or distracted by their own distress about the patient or other family members.

3. To establish a working relationship with health care professionals. Family members often need to identify one person whom they trust and upon whom they rely for information. This task is often complicated in modern hospitals where teams of professionals from various disciplines participate in the care of the patient.

4. To reduce the stressful impact of the illness so that they can attend to other aspects of daily life, for example, other family matters.

5. To maintain a sense of competence and self-worth as members of a family and in the other roles they fulfil in daily life.

6. To recognize potential problems and the means of responding to them especially after the patient is discharged from hospital.

In response to acute illness the family may need to grieve and to restructure itself in the short-term to deal with the redistribution of tasks and roles, that is, responsibility for financial decisions, child-care, assisting with rehabilitation, and deciding which family issues should be deferred while the patient recovers.

Through these activities the family may gain a new sense of itself, and members may discover competencies or resolve pre-existing interpersonal conflicts. On the other hand the patient's own cognitive and emotional reactions to illness and his reciprocal response to those of his wife and other family members may lead to profound distress, fear, helplessness, and inappropriate behaviour which interfere with treatment and may contribute to increased mortality and morbidity. A screening protocol for identifying such families has been proposed (Worby *et al.* 1991).

Family factors in post-infarct rehabilitation

Since the 1950s, clinicians and researchers have reported a high incidence of invalidism in post-infarct patients, most of it due to psychosocial factors. Anxiety, depression, irritability, sleep disturbance, subjective weakness, chest pains, and shortness of breath are commonly reported, resulting in anxiety, over-vigilant concern, and depression in the wife. Marital tensions reflect changes in power-relations, dependency, and concern about death, financial matters, and the family's welfare. Sexual difficulties are common. It has been said that the patient may recover from a coronary but his wife may not. Children too may be adversely affected.

All these difficulties may be evident at the time the patient returns home and may persist for years. A major source of conflict and distress is the differing interpretations made by husband and wife of the information they have been given by medical, nursing, and other staff about the nature of the illness, its treatment, and ways of preventing a recurrence.

Treatment implications

In the light of these observations brief psychotherapy based on supportive and counselling principles conducted by nursing and medical staff in Coronary Care Units during the immediate post-infarct period have been shown to be very effective in reducing the wife's anxiety. This effect appears to persist for many months following discharge from hospital. Similar beneficial effects of brief individual therapy for the patient have been reported but there is a paucity of interactive studies of marital and family intervention in the post-infarct period.

Staff in coronary care and cardiac rehabilitation clinics also recognize the need to avoid giving ambiguous or broadly interpretable advice, such as 'don't overdo it'.

Involvement of the spouse has been demonstrated to be effective in educational programmes aimed at improving patient compliance with rehabilitation, changing risk factors (diet, smoking, exercise, or Type A behaviour) and his compliance with antihypertensive and other medications. The facilitatory role of the spouse in such programmes is affected by her own beliefs about the nature of the illness and about the value of rehabilitation and relapse-prevention programmes. This is particularly important if the patient continues to deny the reality of his condition (Bar-On and Dreman 1987).

CHRONIC PHYSICAL ILLNESS AND THE FAMILY

We will not attempt the daunting task of considering the relevance of family factors in every chronic medical condition. A set of general principles were described in the introductory section and we now focus on some common chronic disorders in greater detail. Obviously, most clinical conditions described in this book may be considered chronic illnesses, and the chapters on alcoholism and dementia are especially relevant to any discussion of the interaction between chronic physical illness and family life.

Family influences in aetiology

Life-events and personality variables have been mooted as causal factors in many chronic diseases. We have already noted conceptual and methodological difficulties that warrant great caution when attributing causal links to correlations between psychosocial variables and disease.

Personality vulnerability and family of origin

Middle-aged men recently diagnosed as suffering from essential hypertension who scored high on Type A ratings report a much higher incidence of childhood trauma (for example early parental death, physical abuse, chronic parental illness, and parental divorce) than do normotensive controls (Ekeberg *et al.* 1990). Likewise, in his famous study of 'the pain-prone patient', Engel (1959) described such a person's childhood as typified by violence, hostility, rejection, and recurrent disappointment, a family life that was so overshadowed by the chronic illness or recurrent pain in a parent or sibling that the patient experienced a sense of continuing neglect except for occasions when he was himself ill or in pain.

Women diagnosed with breast cancer show an excess of so-called Type C personality features, such as self-effacement, conscientious concern for the needs of others, and suppression of their own feelings. In their families of origin they often were the 'over-responsible' parentified child who cared for a sick or depressed parent.

Studies of the interaction between genetic and environmental (especially family) factors in the aetiology of disease represent a sophisticated advance over the older 'either-or' models (Reiss *et al.* 1991). This question is addressed in other chapters of this book, notably in that on schizophrenia.

Genetic factors are known to play a role in diabetes, especially in non-insulin-dependent diabetes. In a study of diabetic twins it was noted that in their young adult years most of the discordant twin pairs lived apart, whereas only half of the concordant pairs did, suggesting the possibility of a shared environmental factor. Other twin studies in which one twin is clinically normal and has a normal glucose tolerance, but also shows evidence of pancreatic islet-cell dysfunction, suggest that environmental (including family) stressors mediate the development of clinical disease in a biologically vulnerable individual. These and other studies of psychosocial factors in diabetes have been reviewed by Helz and Templeton (1990).

Family interaction

In the chapter on eating disorders we summarized the pioneering work of Minuchin and his colleagues who described particular family characteristics associated with a variety of chronic childhood diseases over and above the influence such family factors may have had on treatment compliance. The implication was that family relationships could exert a causal influence on the development and onset of certain chronic diseases. Subsequent workers hold the view that these patterns of family interaction are more important in maintaining than causing illness.

Similar patterns have been observed in families of chronically ill adults, often extending across the generations, so that in each generation spouses are chosen and children raised in ways that ensures that the basic relational rules of the family system remain unchanged. (Wirsching *et al.* 1981). These relational rules are:

(1) conflicts and anger are bad; they mostly lead to misery and must therefore be suppressed under all circumstances;

(2) every member should be available for the other, the family being the only safe refuge in a hostile world;

(3) it is better not to speak about the disease and its consequences (or any other problems) in order not to burden other family members.

Criticisms of this pioneering work have been made on methodological, conceptual, and empirical grounds. The model has been modified in its application to chronic inflammatory bowel disease (Wood *et al.* 1989) and diabetes mellitus (Mengel *et al.* 1992).

Impact of the family on the course of chronic illness

Despite a question about the role of family factors in aetiology there is little doubt that they affect the course and degree of morbidity, as well as enhancing the patient's coping ability and compliance with treatment.

Thus poor metabolic control in insulin-dependent adult diabetics may be caused by psychosocial difficulties, of which major illness in a significant other person is a common cause (Wrigley and Mayou 1991). Poor diabetic control may be due to the distress caused by such life events and may be mediated physiologically, behaviourally, or by the lack of social supports.

Other problems in patients with diabetes such as sexual difficulties may have a primary physiological basis which may lead to secondary interpersonal conflicts and thereby interfere with adequate glycaemic control, thereby augmenting and perpetuating the illness. A reciprocal association exists between pathophysiology and interpersonal relationships which may influence the course of chronic illness.

In patients with chronic pain, disagreement between a couple as to its meaning and management increases the distress of both partners, aggravates the patient's pain experience and hence adds to marital and family distress. This further lowers the patient's threshold for pain, thereby setting up a vicious cycle of symptom maintenance. Factors like the degree of depression in either partner, the level of functional impairment of the patient and the reciprocal effects of patient and spouse on their children's behaviour further complicate the clinical picture. It has also been suggested that in a supportive relationship the spouse's sympathetic response to her husband's pain may perpetuate his pain behaviour (that is, inadvertent positive reinforcement), while at the other extreme, when the spouse has become dissatisfied with the marriage (perhaps because of the impact of the husband's pain), she becomes less supportive thereby increasing his sense of isolation, anger, and lowered self-esteem, which leads to aggravation of the pain (Kerns *et al.* 1990).

Three parameters of family function – strong parental coalition, respect for autonomy in the context of closeness, and affectionate, optimistic interaction correlate with young adults' coping with and adherence to long-term haemodialysis in home and hospital-based settings (Steidl *et al.* 1980); these findings are consistent with the view that family and other social support exert a positive effect on a patient's ability to cope with chronic illness. More controversial is the claim that psychosocial support (not derived exclusively from the family) may influence the duration of remission of patients with advanced breast cancer, in addition to the beneficial effects of such support on the patient's mood and degree of pain (Spiegel *et al.* 1989). The immunological process whereby such psychosocial influences are possibly mediated has been proposed (Levy *et al.* 1990).

Equally controversial, though more counter-intuitive is the finding that the mortality rate in a small group of patients with chronic renal disease receiving haemodialysis increased among those families who scored highest on laboratory based measures of problem-solving ability and family cohesiveness (Reiss *et al.* 1986). Several explanations have been suggested for this unexpected finding. It is possible that different family factors are involved in coping with the course of an illness than those associated with adjustment to the terminal stage of the condition. Alternatively, the family's responses and the experimenters' interpretation of the results of the tests of family functioning may differ according to whether the context is a laboratory or naturalistic setting. The study requires replication.

Meanwhile the consensus in the field is that family characteristics associated with successful coping with illness in the patient include:

(1) a clear separation of generations;

(2) flexibility within and between family roles;

(3) clear and consistent communication;

(4) a tolerance for autonomy and individuation.

However, we must recall that these characteristics are relative to the family's life-cycle stage, transgenerational history, cultural norms, and many other factors. As a consequence, the clinician should take care when deciding whether increased closeness, special understanding, suppression of anxieties or differences of opinion observed in a family with a chronically ill member are signs of 'dysfunction' or whether they represent an adaptive response which enhances coping. A sensitive review of families with an adolescent suffering from thalassemia illustrates this point (Georganda 1988).

Impact of chronic illness on the family

Chronic illness may affect every aspect of family life in a practical, material sense in such matters as financial security, recreational activities, social contacts, career plans, and dietary habits.

Family members individually and collectively may feel guilt, shame, and self-blame. This is particularly likely when the patient is a child or adolescent, when the disease has a genetic mode of transmission, or when the family has particular medical 'myths' or its own historical narrative which accounts for the illness (Hardwick 1989).

Diverse expressions of individual and collective grief occur for the healthy person they once knew, for the relationship they previously had with him, and for the sense of family they once shared. Such reactions may recur around the anniversary of the illness, and at significant periods in the life cycle when the patient fails to take the age-appropriate developmental steps as readily as do his peers or other family members.

The family may become socially isolated, because it considers itself to be 'different' or is indeed stigmatized by society.

Common concerns of care givers of a chronically ill patient include:

1. How much to assist and how much to push the patient to be independent?

2. Anxiety generated by fluctuations in the patient's condition and which of these are related to the disease and which are not? Are all apparent disabilities 'legitimate'?

3. Feeling burdened by the patient's depression and irritability.

4. Will the children develop the disease?

5. Deterioration in communication, especially between the partners when one is ill. Family members may avoid certain subjects which are felt to be emotionally 'explosive' (for example, the disease itself, sex, financial, or other concerns).

6. Exhaustion at taking on the roles and responsibilities relinquished by the sick person; this is often compounded by a reluctance to seek outside help.

7. Can the patient try harder?

8. Have we searched adequately for a cure?

9. Problems with the extended family where everyone may have an idea about what should be done.

Feelings that are experienced by family members, especially the principal care-giver that are inevitable yet difficult for the care-giver to accept include: feeling trapped and wishing to leave; distress at missing recreational, social, and work activities; resenting the patient because of this, or other family members and friends for not being more helpful; feeling burdened at having to make decisions and special arrangements for the patient especially when trying to include him in a family activity; wishing to care for oneself; and embarrassed by the patient in public. Such feelings may be more readily expressed by children and adolescents toward their disabled or chronically ill parent than by the spouse. These responses are typical of the care-givers of patients suffering from multiple sclerosis (Larberg and Cavallo 1984), and their distress may indeed increase as the patient successfully adjusts to his chronic illness. Similar concerns are commonly cited as sources of distress among family members caring for a patient with chronic renal disease, head injury, and post-stroke.

Life-saving technologies in the home such as haemodialysis and respirators may have a stressful impact on the spouse and other family members. While the patient may benefit from the extra technological care and the doctors express satisfaction with the treatment, family members, especially children, may feel unable to express their guilt, resentment, and frustration at the disruption of household arrangements and the increased responsibility they may shoulder in caring for the patient and his technological life-line.

We have already mentioned that a family's experience of illness in one generation, especially in terms of its economic and psychosocial consequences, will strongly influence family response to illness in the next generation. This is especially important in genetically transmitted diseases. For example, in Huntington's Disease, an autosomal dominant disorder that tends to become clinically manifest after the age of 30, the family's previous experience of the disease influences individual coping responses, the tendency either to deny or erroneously detect signs of the illness in themselves or in other members according to a family myth as to who will develop the condition. These beliefs may cause considerable distress and lead to suicide (Kessler and Bloch 1989).

Numerous researchers report that the spouses of chronic pain patients suffer a higher incidence of depression, pain-related states, marital and sexual difficulties than do the spouses of patients suffering from chronic, non-painful conditions. As noted previously, marital and other family difficulties may precede the onset of the condition, and in keeping with one of the models described in Chapter 3, the pain may stabilize the troubled relationship, enabling the marital difficulty to be denied and indeed facilitating an idealized view of what the marriage and family life were like before the painful condition occurred.

Family factors in patient compliance with treatment and rehabilitation

Traditionally, medicine has viewed compliance as the responsibility of the patient and non-compliance as an indicator of personality problems or other psychopathology in the patient. Many clinicians acknowledge that difficulties in patient-doctor relationship may also influence compliance. However, a systems-oriented view of compliance has evolved which includes the patient's family as a cogent factor. Generally, patient compliance and participation in the rehabilitation is enhanced by family support. Most attention has concentrated on the patient's spouse; it seems clear that involving the spouse through education about the illness and its treatment, and offering support for her during periods of distress improves compliance. This has been demonstrated in a wide variety of disorders, including hypertension, diabetes, epilepsy, and chronic renal disease.

The day-to-day experiences of living with chronic illness lead many patients and their families to construct views of it which differ considerably from a biomedical model. Clinicians should be aware that a major source of tension between spouses which may interfere with compliance and rehabilitation is their differing interpretations of the doctor's recommendations and instructions.

Principles of treating the family

'Family-sensitive practice', a felicitous term first applied to working with families of patients suffering from chronic mental illness and traumatic brain damage, describes a range of forms of assistance a family as a group may require (Perlesz *et al*. 1992). These include: education, support groups, networking, advocacy, marital and sexual counselling, and family counselling or family therapy.

Such a spectrum of approaches implies flexibility on the clinician's part and a commitment to 'stay with' the family over the long-haul. For example, in a chronic relapsing or progressive illness such as multiple sclerosis there often are 'crises' or nodal points when the family's mode of coping is disrupted and the family's needs should be reviewed. Such times include the period of diagnosis and the occurrence of specific losses. The latter may be acute or progressive, and include financial, employment, mobility, sexual function, bowel and bladder control, and significant relationships (Larberg and Cavallo 1984).

Family members vary in their tolerance for these changes and will often respond with previously applied patterns of behaviour, regardless of whether these are effective or appropriate under the new circumstances. Members also differ in the way they respond to particular behaviour by the patient, for example, some are angry or rejecting, while others are 'over-protective' and solicitous. The point of view of each relevant family member should be sought and assessed as to how it reciprocally connects to the function of the family as a whole (see Chapter 3).

The clinician should pay particular attention to incomplete, inadequate, or conflicting versions of the illness and its treatment held by different members. While the expression of feelings is an integral part of the grief process (see Chapter 4) direct challenges of life-long ways of coping may be destructive for the patient and the family as a whole.

It is also important that the clinician recognize and acknowledge the family's strengths and resources, as evidenced by its previous dealing with trauma, illness, and loss. These may then be mobilized and modified for the current illness. We previously mentioned the two family constructs, role flexibility and rules governing emotional expressiveness which appear to influence successful family adaptation to chronic illness, and these may be the subject of specific family-oriented interventions (Koch 1985).

A short-term (eight-session) multiple-family group approach has shown encouraging results in enhancing family coping. It combines a psycho-educational model with an exploration of how chronic illness organizes family life in ways which interfere with the desirable balance between stability and growth in the context of family life-cycle tasks. Particular attention is paid to the creation of metaphors which reflect the family's identity, the demands of illness, and the family's emotional expression (Gonzalez *et al*. 1989). Similar success has been claimed by the deployment of another brief family therapy technique, the reflecting team discussion (Griffith *et al*. 1992). At first glance these approaches appear to differ markedly from most models of 'family-sensitive practice' which emphasize the long-term nature of the clinician's availability to the family of a chronically ill person. Indeed, without such a commitment from the clinician it is unlikely that the family could have much confidence in him and would feel discouraged from seeking anything other than biomedical help.

The results of both these models obviously should be replicated and their therapeutic factors identified.

Terminal illness

This subject is considered in Chapter 4.

REFERENCES

Appels, A. (1990). Mental precursors of myocardial infarction. *British Journal of Psychiatry*, **156**, 465–71.

Bar-on, D. and Dreman, S. (1987). When spouses disagree: a predictor of cardiac rehabilitation. *Family Systems Medicine*, **5**, 228–37.

Campbell, T.L. (1986). The family's impact on health – a critical review. *Family Systems Medicine*, **4**, 143–4.

Costell, R.M. and Reiss, D. (1982). The family meets the hospital. *Archives of General Psychiatry*, **39**, 433–8.

Eaker, E.D., Haines, S.G., and FeinPeib, M. (1983). Spouse behaviour and coronary heart disease in men. Prospective results from the Framingham Heart Study. *American Journal of Epidemiology*, **118**, 23–41.

Ekeberg, O., Kjeldsen, S.E., Eide, I., and Leren, P. (1990). Childhood traumas and psychosocial characteristics of 50 year-old men with essential hypertension. *Journal of Psychosomatic Research*, **34**, 643–9.

Engel, G.L. (1959). Psychogenic pain and the 'pain prone' patient. *American Journal of Medicine*, **26**, 899–918.

Engel, G.L. (1968). A life-setting conducive to illness: the giving-up-given-up complex. *Annals of Internal Medicine*, **69**, 293–7.

Fisher, L., Terry, H., and Ransom, D.C. (1990). Advancing a family perspective in health research: models and methods. *Family Process*, **29**, 177–89.

Georganda, E.T. (1988). Thalassemia and the adolescent: an investigation of chronic illness, individuals and systems. *Family Systems Medicine*, **6**, 150–61.

Gonzalez, S., Steinglass, P., and Reiss, D. (1989). Putting the illness in its place: discussion groups for families with chronic medical illnesses. *Family Process*, **28**, 69–87.

Griffith, J.L., Griffith, M.E., Krejmas, N. *et al.* (1992). Reflecting team consultations and their impact upon family therapy for somatic symptoms as coded by structural analysis of social behaviours (SASB). *Family Systems Medicine*, **10**, 53–9.

Hackett, T.P. and Cassem, N.H. (1969). Factors contributing to delay in responding to signs and symptoms of acute myocardial infarction. *American Journal of Cardiology*, **24**, 651–6.

Hanson, C.L., Klesger, R., Ecki, L. *et al.* (1991). Family relations, coping style and cardiovascular risk factors among children and their parents. *Family Systems Medicine*, **8**, 387–98.

Hardwick, P.J. (1989). Families' medical myths. *Journal of Family Therapy*, **11**, 3–27.

Helz, J.W. and Templeton, B. (1990). Evidence of the role of psychosocial factors in diabetes mellitus: a review. *American Journal of Psychiatry*, **147**, 1275–82.

Holmes, J.H. and Rahe, R.H. (1967). The social adjustment scale. *Journal of Psychosomatic Research*, **11**, 213–18.

Jaber, R., Steinhardt, S., and Trilling, J. (1991). Explanatory models of illness: a pilot study. *Family Systems Medicine*, **9**, 39–51.

Kerns, R.D., Haythornthwaite, J., Southwick, S., and Giller, E.L. (1990). The role of marital interaction in chronic pain and depressive symptom severity. *Journal of Psychosomatic Research*, **34**, 401–8.

Kessler, S. and Bloch, M. (1989). Social systems response to Huntington disease. *Family Process*, **28**, 59–68.

Kleinman, A. (1988). *The illness narrative: suffering, healing and the human condition*. Basic Books, New York.

Koch, A. (1985). A strategy for prevention – role flexibility and affective reactivity as factors in family coping. *Family Systems Medicine*, **3**, 70–81.

Larberg, J. and Cavallo, P. (1984). The family reaction. In *Multiple sclerosis – psychological and social aspects* (ed. A.F. Simons), pp. 42–53. Heinemann, London.

Levy, S.M., Herberman, R.D., and Whiteside, T. (1990). Perceived social support and tumour oestrogen–progesterone receptor status as predictors of natural killer cells activity in breast cancer patients. *Psychosomatic Medicine*, **52**, 73–85.

Lipowski, Z.J. (1970). Physical illness, the individual and the coping process. *Psychiatry in Medicine*, **1**, 91–102.

Mayou, R. (1989). Illness behaviour and psychiatry. *General Hospital Psychiatry*, **11**, 307–12.

McDaniel, S.H., Hepworth, J., and Doherty, W.J. (1993). A new prescription for family healthcare. *The Family Therapy Networker*, **17**, 18–29.

Medalie, J.H. and Goldbourt, U. (1976). Angina pectoris in 10 thousand men. Psychosocial and other risk factors evidenced by a multivariate analysis of a five year incidence study. *American Journal of Medicine*, **60**, 910–21.

Mengel, M.B. (1987). Physician ineffectiveness due to family-of-origin issues. *Family Systems Medicine*, **5**, 176–90.

Mengel, M.B., Blackett, P.R., Lawler, M.K. *et al.* (1992). Cardiovascular and neuroendocrine responsiveness in diabetic adolescents within a family context: association with poor diabetic control and dysfunctional family dynamics. *Family Systems Medicine*, **10**, 5–33.

Miller, P., Dean, C., Ingham, J.G. *et al.* (1986). The epidemiology of the life events with some reflections on the concept of independence. *British Journal of Psychiatry*, **148**, 686–96.

Patterson, J.M. (1988). Families experiencing stress. The family adjustment and adaptation response model. *Family Systems Medicine*, **5**, 202–37.

Patterson, J.M. (1990). Family health research in the 1980's: a family scientist's perspective. *Family Systems Medicine*, **8**, 421–34.

Perlesz, A., Furlong, M., and McLachlan, D. (1992). Family work and acquired brain damage. *Australian and New Zealand Journal of Family Therapy*, **13**, 145–53.

Power, P. (1991). Family coping with chronic illness and rehabilitation. In *Handbook of studies of general hospital psychiatry*, (ed. F. Judd, G.D. Burrows, and D. Lipsett). Elsevier, Amsterdam.

Price, V.A. (1988). Research and clinical issues in treating type A behaviour. In *Type A behaviour: research therapy and interaction* (ed. B.K. Houston and C.R. Snyder), pp. 275–311. Wiley, New York.

Reich, P., de Silva, R.A., Lown, B. *et al.* (1981). Acute psychological disturbance preceding life-threatening arrhythmias. *Journal of the American Medical Association*, **246**, 233–9.

Reiss, D., Gonzalez, S., and Kramer, N. (1986). Family process, chronic illness and death. *Archives of General Psychiatry*, **43**, 795–807.

Reiss, D., Plomin, R. and Hetherington, E.M. (1991). Genetics and psychiatry: an unheralded window on the environment. *American Journal of Psychiatry*, **148**, 283–91.

Rolland, J.S. (1984). Toward a psychosocial typology of a chronic and life-threatening illness. *Family Systems Medicine*, **2**, 245–63.

Rolland, J.S. (1987). Chronic illness and the life-cycle: a conceptual framework. *Family Process*, **26**, 203–21.

Rozanski, A., Bainey, N., Krantz, D.S. *et al.* (1988). Mental stress and induction of silent myocardial ischaemia in patients with coronary artery disease. *New England Journal of Medicine*, **318**, 1005–12.

Spiegel, D., Bloom, J.R., Kramer, H.C. *et al.* (1989). Effects of psychosocial treatment on survival of patients with metastatic breast cancer. *Lancet*, **2**, 888–91.

Steidl, J.H., Finkelstein, F.O., Wexler, J.P. *et al.* (1980). Medical condition, adherence to treatment regimens and family functioning. *Archives of General Psychiatry*, **37**, 1025–7.

Steinglass, P. and Horan, M.E. (1988). Families and chronic medical illness. In *Chronic disorders and the family* (ed. F. Walsh and C. Anderson), pp. 127–42. Haworth Press, New York.

Strain, J.J. (1978). *Psychological interventions in medical practice*. Appleton-Century-Crofts, New York.

Turk, D.C., Rudy, T.E., and Flor, H. (1985). Why a family perspective for pain? In *The family and chronic pain* (ed. R. Roy). Special issue of *International Journal of Family Therapy*, **7**, (supplement 4) 223–34.

Wirsching, M., Stierlin, H., Weber, G., and Wirsching, B. (1981). Family therapy with physically ill patients. In *Psychosomatic factors in chronic illness* (ed. K. Achte and A. Pakaslahti) *Psychiatria Fennica Supplementum* 1981. Foundation for Psychiatric Research in Finland, Helsinki.

Wood, B., Watkins, J.B., Boyle, J.T. *et al.* (1989). The 'psychosomatic family' model: an empirical and theoretical analysis. *Family Process*, **28**, 399–417.

Worby, C.M., Altrocchi, J., Veatch, T.L., and Crosby, R. (1991). Early identification of symptomatic post MI families. *Family Systems Medicine*, **9**, 127–35.

Wrigley, M. and Mayou, R. (1991). Psychosocial factors and admission for poor glycaemic control: a study of psychological and social factors in poorly-controlled insulin-dependent diabetic patients. *Journal of Psychosomatic Research*, **35**, 335–43.

Wynne, L.C., Shields, C.G., and Sirkin, M.I. (1992). Illness, family theory and family therapy. *Family Process*, **31**, 3–18.

9

Alcohol and drug abuse

ALCOHOLISM

To understand the problem of alcohol abuse is to know all of psychiatry, in its biological, psychological, and sociocultural aspects. This chapter concentrates on one area of this vast field, namely the connection between alcohol abuse and family factors. The terms heavy drinking, problem drinking, alcohol abuse, and alcoholism will be used interchangeably for convenience although a specialized literature exists regarding the distinctions between these terms and their implications (Hasin *et al.* 1990).

The co-existence of alcoholism and most forms of psychiatric disorders (that is, co-morbidity) is so common that indicators of alcohol abuse by the patient and by other family members should always be sought in a psychiatric diagnostic interview, regardless of the nature of the presenting problem. Alcohol and drug abuse in particular also often co-exist in certain patient groups.

Alcoholism may develop in response to a primary psychiatric condition, that is, alcohol or drug abuse is an attempt to self-medicate and relieve the distress of painful thoughts and feelings (Khatzian 1985). In other cases alcoholism may develop independently, with a diverse range of psychiatric states resulting as specific consequences of adverse biological, psychological, and social effects of alcohol abuse.

Psychiatric states in the family of origin

The biological families of alcoholic patients show an increased prevalence of a variety of forms of psychopathology, including alcoholism, major depressive illness, and antisocial personality disorder (Hesselbrock 1986). Patients with substance abuse, including opiates and stimulant drugs, have an increased prevalence of alcoholism in their families of origin.

Families also show an increased risk of obesity, eating disorders, and gambling. Social learning theory links alcohol abuse and these disorders in families to a failure to learn appropriate controls over behaviour during socialization in the family; psychoanalytic theorists view the association between alcoholism and these family difficulties as implying a defective sense of self arising from family relationships.

The family dynamics of the alcoholic patient often are discussed in terms of the relationship with the wife and children, whereas those of the drug abuser usually are viewed in terms of his relationship with his family of origin (that is, parents and siblings). These two 'levels' of family dynamics are interrelated (Bowen 1974), and will be discussed below in the section on substance abuse.

Genetic factors

Rates of alcoholism and drug abuse are significantly greater for mono-zygotic than for dizygotic twins. Adoption studies of children of parents with alcoholism and drug abuse confirm the likely effect of a genetic factor. This applies particularly to the sons of an alcoholic father but is less clear for daughters.

Research has classified two types of alcoholism (Dinwiddie and Cloninger 1989):

1. Type 1, milieu – limited, has its onset relatively late, over 30 years of age, and occurs in a person who shows anxiety and marked passive-dependent personality traits. Socially, the person is shy, sensitive to social cues, pessimistic, and cautious. Men and women are equally represented in this group. Alcohol is used as an anxiolytic; tolerance to and dependence upon alcohol develop rapidly. Either or both biological parents has a history of alcoholism but not of criminality. A genetic element has been postulated in this transgenerational transmission which can be influenced to a great extent by environmental (for example, family) factors, so that the genetic effect is either increased or reduced.

2. Type 2, male-limited, occurs mostly in men, has a relatively early age of onset under 20 years, with binge drinking prominent. Many of these men and their fathers have features of antisocial personality features. Genetic transmission is from father to son, and is not modified by environmental factors. Daughters of Type 2-alcoholic fathers do not have an increased rate of alcoholism, but seem at risk to suffer somatoform disorders.

This categorization is still tentative; it may describe two subtypes of alcoholism that conform to a medical disease model of alcoholism, in contrast to other forms of alcoholism.

Alcoholism, depressive illness, and antisocial personality disorder occur together in the one individual with an increased frequency; a similar association has been noted in the biological families of alcoholic patients.

However, since these three clinical entities are probably genetically distinct the nature of their association remains unclear. Furthermore, it is not clear just what is being inherited. Research suggests that one subgroup of alcoholic patients may inherit a tendency for alcoholism to progress in severity, a process for which a modifier gene at the D2 dopamine receptor locus has been implicated and a similar factor may predispose to other substance abuse (Uhl *et al.* 1992).

Social learning theory

Studies applying social learning models suggest that what is transmitted across the generations are behavioural responses to the ingestion of alcohol, or particular behavioural styles of drinking. In any given family, a child's pattern of drinking in adulthood may be particularly influenced by one or other parent's drinking habits which may reflect underlying family loyalties and identifications. These patterns may be further modified in both healthy and unhealthy ways during adolescence and young adulthood when peer pressures may outweigh the influence of parents on drinking and drug taking patterns.

Many people who are at risk of alcoholism because of apparent genetic, personality, family environment, and sociocultural vulnerability factors do not become problem drinkers. The various ways this occurs have yet to be understood fully, though family relationships appear crucial (Zucker and Lisansky–Gomberg 1986).

Effect on children

Children of an alcoholic parent are prone to anxiety and depression, conduct disorders, hyperactivity, delinquency and truancy, cognitive impairment and learning difficulties, poor peer relations, somatic problems, suicide, and are at risk for physical and sexual abuse, alcoholism and substance abuse.

ALCOHOLISM IN WOMEN

It is only since the 1970s that the problem of alcoholism in women has begun to receive appropriate attention in the clinical literature. Several studies have noted a growing prevalence of alcoholism and alcohol-related problems over this time. The reasons for this are unclear – it may reflect women's rejection of their conventional sex-role stereotype in society, increased access to alcohol, or a greater willingness by women to disclose their problems, especially to agencies that deal specifically with women's health. It is clear however that the prevalence of alcoholism among women in Western society is much greater than the estimates of thirty or forty years ago.

The childhood of the woman with alcoholism, like that of her male counterpart, is typically described as traumatic, disruptive, and emotionally deprived. Parental death, divorce, or desertion are common. She is much more likely to be a victim of incest and other forms of childhood sexual and physical abuse than a non-alcoholic woman (Hurley 1991).

The alcoholic woman's adolescence reveals concerns about her self-concept as a person and as a woman. Two patterns of adolescent behaviour have been described (Sandmaier 1980). In one, the young girl rebels against the feminine norms of her family and society; she becomes defiant of authority at home, at school, and in the community, and often is sexually precocious. In the other pattern she appears to over-value and to conform rigidly to a traditional female role as a way of solving her identity conflict.

The onset of heavy drinking often follows a traumatic event in adult life, such as infidelity by her husband, desertion, divorce, death of a significant family member, a child leaving home, or gynaecological problems. Previously effective, albeit fragile, ways of coping with self doubts are overwhelmed by these family and personal stressors and her use of alcohol to soothe her distress is potentiated by the hostile responses of husband, family, and friends to her drinking.

Depressive illness, eating disorders especially binge-eating and bulimia nervosa, and multiple drug abuse co-exist with alcoholism in women, which complicates both family dynamics and treatment approaches. The woman who drinks heavily is also more likely to be a victim of physical, sexual, and emotional abuse and violence, both in her home and in society.

The divorce rate among women with alcoholism who have non-alcoholic husbands is significantly greater than that of alcoholic men with non-alcoholic wives, suggesting that men are less tolerant of their wife's drinking than vice-versa. Since the impact of divorce on women is generally more severe than on men, especially in terms of housing, job prospects, and income, the divorced alcoholic woman is likely to be in a very vulnerable position, both psychologically and socio-economically. This is even more the case if she has responsibility for the care of children.

About 50 per cent of cases where a woman is the identified problem drinker have a husband who is also a problem drinker. However such women are more likely to be stigmatized, than are their alcoholic husbands, by society at large and by their own families for their drinking and alleged or actual neglect of their children.

Pregnant women who abuse alcohol, tobacco, marijuana, opiates, and cocaine increase their risk of having babies who suffer from a variety of intra-uterine and post-natal physical and developmental defects (McCance-Katz 1991). The best known of these is the fetal alcohol syndrome, which occurs in 1–2 per 1000 live births in the USA, while less severe variants of the syndrome have a much greater incidence. The neurological, behavioural, and psychiatric disturbances displayed by these babies and young children augurs a major health problem for society, as do the long-term consequences of disturbed parent-child interactions which have been observed in chemically dependent women (Davis 1990).

We now focus in detail on the male alcoholic patient and his family.

ALCOHOLISM IN MEN

Marital and family issues in the male alcoholic patient

Three models have been advanced to account for the marriage and family life of the alcoholic patient: the wife's disturbed personality, a stress-response model, and systems-oriented model.

The wife's disturbed personality – this focus derives from the psychoanalytic therapy of the wives of alcoholic men during the 1950s. The woman often came from a family where her own father was a heavy drinker and was perceived as harsh, punitive, or emotionally absent. The woman went on to marry a man who resembled her father, or conversely a passive, dependent person whom she tried to control in order to avoid a repetition of the traumas her own father had inflicted upon herself and her mother.

This view implies that the woman unconsciously chose a husband with the potential to become an alcoholic. This may have promoted an attitude among clinicians which is unsympathetic to the wife of the alcoholic male. However, the reported features of the wife might be more reasonably interpreted as a response to the manifold problems that her alcoholic husband creates for her and her children, that is, the stress–response model.

The pioneering work of Jackson (1954) and subsequent research led to a *stress–response model* which described the progressive effects of the alcoholic man on family life. The model identified a number of stages, each marked by specific coping strategies adopted by the wife and children to deal with the impact of the drinking husband and father on their lives. The model is described in some detail because of its importance:

1. Efforts to avoid, deny, or minimize the problem. This includes making excuses for the husband, protecting him from the consequences of his drinking, and emphasizing his good qualities which they perceive when

he has not been drinking. The wife however feels increasingly rejected and unloved, increasingly angry and frustrated as her attempts to help her husband fail.

The children sense tension between the parents. They soon learn not to ask questions about their father's drinking or to accept the unsatisfactory answers they receive if they do enquire about the welfare of either parent. A similar pattern of denial and rationalization occurs among members of the extended family, friends, workmates, and employers.

If help is sought at this point, which is unlikely, it is usually the wife, who presents with any one or more of a variety of problems (see below), but the alcohol abuse is rarely mentioned.

2. The family increasingly isolates itself from contact with others. The wife's isolation leads her to focus attention on ways of discouraging her husband's drinking: the more preoccupied she becomes with this the more he appears determined to defy her. She then becomes increasingly angry, self-critical and self-pitying, or feels a failure.

The children witness this hostility between the parents, and are themselves subject to inconsistency and unpredictability in the parents' mood and behaviour which may range from excessive solicitousness to inappropriate anger, and both verbal and physical abuse.

Extended family and friends are alienated by the husband's behaviour or are avoided by wife and children because of their own embarrassment or anxiety. Alongside this the husband may form new 'friends', mostly 'pub friendships', who generally aggravate the problem.

3. Family disorganization. Having failed to deny or eliminate the drinking problem recurrent family crises occur which lead to further drinking and thus to further crises.

Often one or more children are 'triangulated', and feel caught in the parental crossfire. The children's concern about the effects of their father's drinking on his health and state of mind may be counterbalanced by a concern for the welfare of their mother.

The stereotype of the 'neurotic, controlling' wife mentioned above, conceivably derives from her seeking help at this stage. The husband may still present himself plausibly, as a competent person who occasionally 'has one or two too many', but is victimized by an intolerant spouse. Since the wife is indeed bitter towards him, as well as riddled with self-doubt and guilt as to her role in his drinking, and may also be so fearful of repeating her mother's 'failures' as wife and mother, she may convey the impression of having considerable 'psychopathology' and of being the prime cause of her husband's drinking problem.

Professional help may be sought at this stage by the wife for herself. She may omit direct reference to her husband's drinking or may understate its severity. Alternatively, the children may present with behavioural problems. Outside sources (for example, police, debt-collector, child-welfare, or other agencies) may persuade the husband to seek help.

4. The attempt at reorganization, – the wife takes over all the essential decision making and financial arrangements, her task is one of safeguarding her own interests and those of her children. She no longer denies that her husband has a serious drinking problem. She may join Al-Anon, and other community groups to support her in an effort to provide for the family.

The children may seek to reconcile the parents, or they may accept that their father is the way he is and resign themselves to an unsatisfactory and at times traumatic home life.

5. The attempt at reorganization may fail, either because the husband's behaviour is increasingly destructive or because the wife's strategy to lead a separate life while remaining formally in the marriage proves unworkable. She may then decide to separate from her husband, but experiences considerable guilt about this decision, and be pressured by her children, family, and friends to remain in the marriage, especially if the seriousness of the drinking problem is not fully appreciated.

6. In the event of separation, wife and children reorganize family life without him.

7. Alternatively recovery of the family may succeed if the alcoholic father seeks help.

This sequence is not invariant and a particular stage may recur several times. Furthermore, stages are not clearly delineated from one another. Since Jackson's original research was based on clients attending Al-Anon the findings may not be generalizable. Moreover, the model may not apply to families in which the wife is the alcoholic parent, or where both spouses abuse alcohol.

In the 1970s investigators studied the coping-response style of the wife of the male alcoholic. The relationship between her personality, the stress model and coping-response pattern led to a complex model of her influence in determining the course of the husband's alcoholism (Moos *et al.* 1982). This work has usefully widened the scope of enquiry beyond questions of individual psychopathology in the alcohol abuser or spouse, and points to a systems model.

A *systems view* emphasizes that regardless of the original causal factors, alcoholism may be maintained by the responses of significant others, particularly the family, while the drinking in turn organizes the family into a particular pattern of relationships. This picture may also apply to the extended family and other social groups.

Kaufman and his colleagues (1984) described four family constellations based on their observation of family dynamics.

1. The functional family – the level of conflict and disruption is minimal, with the problem that of a highly regarded person dealing with personal difficulties through the excessive use of alcohol. The family is satisfied with relationships between its members and defines the drinking problem as specific to the drinker.

2. The neurotically, enmeshed family – severe marital and other inter-personal problems prevail. Heavy drinking by either or both spouses reflects their disillusionment and resentment about their relationship. Conversely, drinking maintains a pattern of chronic conflict and threat-ened separation, alternating with periodic reconciliation between the spouses. Children are often involved by serving as co-parent or by manifesting problems in such a way that unites the parents again, albeit temporarily.

3. The disintegrated family – the family has fallen apart because of the drinker's abusive, violent, and neglectful behaviour. Typically, the husband has underlying personality difficulties; the wife, after years of resentment, leaves him, taking the children. He maintains infrequent contact with his children or with his extended family, most of whom have ostracized him. While it is clear that the husband warrants help in his own right, there are family issues to be clarified, especially the kind of relationship he will continue to have with the family.

4. The absent family – there is no functioning family system or members who could be recruited to relate to the patient. He may survive only on the margins of society, requiring institutional support.

The family system in the life-cycle context

Steinglass and his colleagues (1980) conducted noteworthy studies on patterns of relationships between the alcohol abuser and family in different phases of the life cycle. They noted that periods of intoxication, while undoubtedly destructive, nevertheless allowed for important interactions between family members which did not occur during periods of sobriety. An example is the relationship between two brothers which is usually tense and rivalrous, but when the alcoholic brother becomes intoxicated the relationship changes into one where both express affection for one another.

Furthermore, Steinglass and his colleagues observed that stability or instability in family relationships did not correspond directly with periods of sobriety or intoxication. Rather, there was a complex interaction between degree of sobriety of the alcoholic, developmental demands of the family life cycle, and responses of family members to the drinker whether drunk or sober.

Thus alcohol, the research group postulates, serves two functions in a family: it may indicate individual or interpersonal distress (that is, the traditional view) or it may serve a homeostatic or emotional regulatory function (that is, systems view).

The Steinglass group describes 'dry', 'wet', and transitional phases in the life cycle of the alcoholic person and his family, each phase reflecting an alteration in the drinking pattern, with a corresponding change in family relationships and with phase-specific treatment implications.

The family system may be considered 'dry' when the person is not actively drinking, but it may still be an alcoholic family in that family life is organized around the concern of 'relapse'. In a non-alcoholic system family life is no longer regulated by the person's drinking or abstinence.

As each phase, 'wet' or 'dry' may be either 'stable' or 'unstable', four possible patterns emerge:

1. Stable-wet, alcoholic – in which the family's life is organized around alcohol, this possibly enduring for decades.

2. Stable-dry, alcoholic – the patient no longer drinks and maintains his abstinence. However family life is centred around the fear of relapse, which influences all family activities (for example, holidays, social life). Membership of AA, Al-Anon, and Al-Ateen is a serious commitment, and some family members develop careers in alcoholism and substance-abuse treatment centres.

3. Stable-dry, non-alcoholic – alcoholism is no longer a central issue for the family. Alcohol is not present, either materially or as an emotional factor.

4. Stable, controlled-drinking, non-alcoholic – although still controversial, a return to controlled or social drinking is possible for some individuals.

The number of patients in this pioneering study was small and it concentrated mainly on interactions between adult family members. However partial confirmation of the model was provided by Jacob and Krahn (1988) who studied the effects of alcohol on marital interaction in clinically depressed, non-alcoholic families compared with depressed, alcoholic families.

However, Steinglass' study did not allow for the possibility that the pattern of drinking itself may influence marital interaction and family organization. Thus steady, daily drinking may be more readily incorporated into family life and serve the homeostatic function postulated by Steinglass, whereas binge drinking (for example, on weekends) may be more disruptive (Jacob 1992). The integration of the systems and stress–response models may retain the advantages and reduce the limitations of each model, both conceptually and practically (Orford 1985).

CLINICAL PRESENTATION

As mentioned earlier the range of clinical presentations of alcoholism and the variety of clinical problems in which alcohol abuse is a complicating factor are so vast as to warrant diligent inquiry about the extent and pattern of alcohol and psycho-active drug use by every patient.

Specific inquiry should be made of the alcoholic patient's wife's use of analgesic and psychotropic medication in addition to that of alcohol. Psychosomatic symptoms, anxiety and depressive features, signs suggestive of physical injury, vague but persistent concerns about her children, and excessive guilt over their seemingly trivial problems or illnesses may all be part of the presentation of the wife of an alcoholic man, or of a woman who is herself abusing alcohol and other drugs. Physical violence, child sexual abuse, and coercive sex in marriage are recurrent traumas for the family of the alcoholic person (see Chapter 13).

Psychiatric evaluation of children or adolescents, regardless of the presenting problem, usually includes assessment of the parental relationship and that between each parent and each child. Specific inquiry is made about the effect of alcohol on such relationships. Children are often reluctant to disclose this information, and often feel deeply ashamed and guilty about parental drinking. Children will avoid close friendships, particularly since these will require them to reciprocate invitations to a friend's home or to exchange information about parents.

Paradoxically, the presenting problem may be the child or adolescent's relationship with the non-alcoholic parent. The intensity of such a relationship may distract the clinician from recognizing the alcoholic parent in the family. Over the years a mutually supportive relationship often develops between the non-alcoholic parent and the child, about which the latter feels deeply ambivalent or guilty when he (or she) tries to separate and individuate from the family.

By contrast, the neglectful attitude of the alcoholic parent may allow the youngster a welcome degree of freedom, so that the non-alcoholic parent feels betrayed by her child.

ASSESSMENT

The clinician should seek information from the patient about his use of alcohol and other psycho-active substances in a way which is neither judgemental nor allows him to avoid cooperating.

Details of the clinical assessment will be found in most standard textbooks of psychiatry. Evaluation of the alcoholic person's family follows the principles outlined in Chapter 3.

Once the problem of alcohol abuse has been raised, whether by the clinician, family members, friends, or workmates or a social agency, it is a serious error for the clinician to minimize its importance. Indeed, the clinician should neither collude with the patient in denying the problem nor dismiss the concern and courage of the family member who has divulged, notwithstanding powerful family and societal resistance, that the family has a problem with alcohol abuse.

Where the patient admits that he has an alcohol problem the clinician should ask 'why now?' Why is the patient or spouse presenting for help now? He may be pushed by his wife's ultimatum that either he seek treatment or she will divorce him. In this event the wife should also be asked 'why now?' about her ultimatum. Inquiry should be made about her fears for her own safety and that of her children.

Once the clinician ascertains the existence of a drinking problem he should inform the patient clearly, but non-judgementally. The spouse is informed too in the patient's presence. Whether other members, especially young children, are also told depends on several factors, including the extent to which their lives are affected, the patient's wishes, and the clinician's treatment plan.

TREATMENT

A range of physical, psychological, and social treatments, in individual, marital, family, and group settings, are used in the management of the alcoholic patient. No single method can claim to work optimally for all patients at all times; a variety of treatments in differing combinations is usually necessary at specific stages of the therapeutic programme.

Admission to hospital may be indicated for detoxification (alcohol withdrawal). Other indications include suicidal risk, violent propensities, underlying psychiatric problems (for example, mood or anxiety disorder, paranoid or jealous state, organic brain syndrome), poor physical health, and severe family conflicts, which include risk of violence or of self-destructive behaviour on the part of the patient or other family members. Sometimes the severity of these factors may not be determined until the patient has been hospitalized and ceases drinking. The patient's admission itself may provoke a family crisis. Accordingly some specialist clinics have facilities for admitting the whole family to a residential facility for a period of intensive assessment and treatment.

The clinician may help introduce the patient and family members to *self-help groups* which are available in the community and in many hospital settings. The most popular is AA for the alcoholic patient, Al-Anon for the spouse, and Al-Ateen for the teenagers and older children. Similar groups exist for those who have abused substances including narcotics, cocaine, and psychotropic drugs (for example, benzodiazepines). Referral to these groups does not mark the end of the clinician's responsibility. Not all patients and their families find the AA approach acceptable, and even if they do, important issues still need to be addressed by the clinician with the patient and other members. For example, many self-help groups emphasize personal responsibility, encourage family members to detach themselves from the drinker rather than trying to change him, and promote supportive contacts with other group members. The clinician can help the patient and family members to reconcile the individually-oriented aims of the self-help group with their wish to promote family cohesion – albeit different from that which existed previously.

Furthermore some forms of group therapy zealously advocate the 'honest sharing' of feelings. This may lead to confrontation between family members. Since marital cohesion is an important factor in treatment outcome, such confrontation, if destructive, may undermine the treatment (Orford *et al.* 1976).

Many patients and their families who attend AA and Al-Anon fit Steinglass' description of a family system that is dry (that is, abstinent) but 'alcoholic' in the sense that family life is still organized around alcohol avoidance and combating the threat of its possible recurrence. Assisting the family to shift to a 'non-alcoholic' system (that is, where they are not organized around the risk of relapse of alcoholism) requires individual, marital, and family therapy of a type not usually available in AA (see below). Despite these differences family systems and traditional AA-based treatments are compatible (de Maio 1989).

Marital and family therapy is useful in: (a) promoting the alcoholic patient's willingness to seek and remain in treatment; (b) increasing the effectiveness of treatment which aims to reduce the amount of drinking or to promote abstinence; and (c) maintaining remission.

Considerable research has demonstrated the use of *behavioral marital therapy* (BMT) in achieving these aims (O'Farrell 1992). Various methods are applied including contingent reinforcement by the spouse, psycho-educational approaches providing information about alcohol's effects, training in family communication, problem-solving, and recognizing contextual antecedents to heavy drinking. Learning and rehearsing alternative approaches to domestic tension and to the urge to drink are practical and cognitive skills which are taught in either a marital or couples' group-context.

However, the superiority of BMT over individual cognitive-behavioural treatment of the alcoholic appears to decline within a year of the end of the treatment programme. This has led to 'booster' or relapse prevention BMT, which in addition to the aforementioned techniques, teaches the couple to recognize early signs of relapse and its precipitants, as well as addressing problems in the marital relationship. Increasingly sophisticated research methods may help identify those patients who may benefit more from individual rather than marital therapy or vice versa.

Concurrent *marital group therapy* based on psychodynamic principles for male alcoholics and their wives (that is, the male alcoholics in one group and their wives in a separate group) was first described by Gliedman and his colleagues in 1956. Although the number of patients was small, the results had interesting applications for the treatment of alcoholism specifically, as well as for psychotherapy in general. While the reduction in drinking behaviour in the alcoholic patients was slight, there appeared to be quite a significant improvement in other psychiatric symptoms (for example, depression, irritability), and a general improvement in reported marital satisfaction by both the husbands and their wives (Gliedman *et al.* 1956). More extensive studies of marital group therapy supported these conclusions. *Multiple couples group therapy*, that is, husbands and wives in the same group setting, has become a popular therapeutic approach to alcoholism, though clinicians vary in the emphasis they give to the group process, to family interactions, or to techniques which attempt to combine both family and group dynamics (Steinglass 1979). Generally, conjoint group therapy with married couples has replaced concurrent group therapy for husbands and wives. Unlike the previously described BMT, the emphasis in this form of therapy is on marital relationships, the expression of feelings, and communication in the marriage rather than on the patterns and amount of alcohol consumption *per se*.

While a *family systems* approach to conceptualizing some of the causal and problem maintenance factors in alcoholism has been enthusiastically accepted by many authorities in the field, the place of formal family therapy has not been adequately evaluated (O'Farrell 1992). Most 'family' therapy continues to be with couples within a systems perspective. The specific treatment issues raised by the inclusion of children in the therapy have not been resolved, nor has the question of how to provide individual therapy for a child or other family members while still adhering to a systems viewpoint. However, some promising avenues of family therapy are evolving (Galanter 1989).

There are also reports of successful family treatments conducted in community-contexts using techniques to increase vocational, social, and recreational skills, and problem-focused support for legal, housing, and financial difficulties (Hunt and Azrin 1973; Davis and Hagood 1979).

The *social network* of the patient includes family, friends, and workmates. It has been found to be a powerful source of support for the patient, and to facilitate abstinence from alcohol or drugs, providing members of that social network do not drink alcohol or use drugs with the patient. Recruiting such people may itself form part of the treatment programme, as the patient has to learn communication skills, appropriate assertiveness, and perhaps deal with unresolved conflicts between himself and his family or friends.

In summary, three broadly defined, overlapping, family-oriented treatment approaches exist (Orford 1990):

(1) those focusing on patterns of drinking, that is, psycho-educational and cognitive-behavioural techniques;

(2) those focusing on interpersonal needs and conflicts;

(3) those focusing on wider contextual and social support problems.

The indications for and comparative efficacy of these approaches are undergoing evaluation.

Relapse factors

As the patient and family members are successful in their efforts to overcome the alcohol abuse new problems may emerge which may lead to relapse. Such problems include:

1. Family members may fear that upsetting the patient will precipitate a 'relapse' and so go to great lengths to avoid upsetting him even to the extent of not disagreeing with him, or complying with his every request. This may lead to new tensions in family life, from which the patient seeks relief through alcohol once again. This is an example of the family's response to an improvement in the patient's condition unwittingly encouraging relapse (see Chapter 3).

2. Personal and interpersonal difficulties which were previously avoided or which were attributed to the drinking problem now become more evident.

3. The patient's unfamiliarity with the new roles which he tries to fulfil in his family, and others' expectations of him now that he is 'better'.

4. Other family members may have to change their attitudes and habitual ways of dealing with the patient and with one another. Thus, the wife may need to learn anew to discuss family decisions with her husband instead of solving all problems on her own or consulting with the oldest child. Likewise, the teenage child who has grown used to acting as a surrogate father to his younger siblings and as a confidant to his mother, may find that his father is now trying to fulfil these roles. The wife may then feel caught in a conflict of loyalties.

5. The increased social contacts that the family may develop, often after years of social isolation, tyranny, and shame. The children begin to experiment with bringing friends home, and the parents, as individuals and as a couple, may cautiously resume contact with extended family and old friends.

6. Divorce – sometimes the patient or his wife may decide to separate from the family. As described above, Steinglass has identified the transitional points in the life cycle where such decisions are likely to occur.

7. An older child may decide to leave the family home. This may be experienced by one or both parents as a severe loss, and may arouse guilt feelings about the alcoholism and other matters. Often this is about their own separation from their parents or the death of a sibling.

The young person's decision to leave the family home should be assessed as to whether it is an age-appropriate thing to do, and whether the manner of doing it indicates some problem. This is an important form of preventative psychiatry, since many young people from alcoholic families who leave home prematurely or inappropriately are victims of various forms of abuse and deprivation, and are at risk of developing alcoholism, other forms of substance abuse and a range of other psychopathology. Furthermore, the young person who leaves home in this way is likely to be unemployed, and is therefore at risk of becoming homeless, exploited, and alienated from society.

If the patient refuses help

A common difficulty is the patient with the drinking problem who refuses to seek help for himself or to attend any family sessions. The wife should be offered some individual therapy for herself, and this may be combined with some family meetings for herself and her children. It may also be appropriate to invite extended family members, close friends, and other significant members of the social network to some of these meetings. It is important for the clinician not to condemn the absent drinker, since some members of his family, especially one or more of the children, may see him as a victim or feel very concerned for him. Differences of opinion as to how the wife should respond to him, including whether or not she should leave him, may reflect unresolved transgenerational issues between the wife and her own parents.

This so-called 'unilateral therapy' may help to:

(a) assess the risks of physical, sexual, and emotional abuse or neglect to the wife and the children;

(b) change, albeit within appropriate limits, the wife's responses to her husband which appear to trigger or aggravate his drinking;

(c) support the wife's efforts to lead a more satisfying life (Thomas and Santa 1982).

PREVENTION

The family which does not allow its life to be dominated by their alcoholic member (that is, a family with an alcoholic member rather than an alcoholic family), provides an environment where the children are less likely to develop alcoholism in their adult years. Practical interventions have evolved which encourage the family (the mother and children) to develop routines like mealtimes, bedtime, weekend, and holiday activities which occur in a regular and predictable way, and which are 'immune' to the disruptive behaviours of the alcoholic parent. Furthermore if such patterns are replicated in the children's marriage, their risk of alcoholism is reduced further (Wolin *et al.* 1980).

DRUG ABUSE

As mentioned earlier, family relationships of the alcoholic patient are best described as those of an adult in a marriage and the relationship he and his wife have with their children, (that is, the family of procreation). The family relationships of the drug abuser (typically opiates) are generally described in terms of an adolescent attempting to separate from his parents (that is, the family of origin). The transgenerational model of family functioning and therapy recognizes that the family-of-origin and the family-of-procreation models are closely related (Bowen 1974).

However, both family patterns may appear concurrently or the pattern may not reflect the chronological age of the patient. Thus the adolescent pattern of substance abuse may continue into early or middle adulthood in cases of either alcoholism or heroin addiction. Although not universally agreed upon, emphasis is commonly given in terms of developmental, diagnostic, and therapeutic considerations to the adolescent pattern before attending to spousal or family-of-procreation issues (Levine 1985).

In a study (Ziegler-Driscoll 1979) comparing patients with drug abuse and alcohol abuse many similarities were found, including a similar proportion of families of origin and families of procreation despite the age differential of the two clinical groups, a high rate of alcoholism in other family members and similar patterns of family dynamics.

Early observers concluded that family dynamics permitted the adolescent substance abuser to avoid age-appropriate developmental tasks. More recent research suggests he often actually has attempted these tasks, but prematurely or precociously, only to fail. As a result and against a background of developmental difficulties, he seeks solace in substance abuse (Newcomb and Bentler 1989).

Family of origin

The use of alcohol by parents are strong predictors of adolescent drug and alcohol abuse (Stanton 1979). Family interaction in adolescents with heroin abuse differs for males and females (Stanton 1979). In the families of origin of the male addict the mother is described as indulgent, over-protective, and permissive toward the patient. The father is described as peripheral, weak, or absent, often with a drinking problem himself. By contrast, female abusers describe their fathers to be sexually aggressive, intrusive, and inept. The patients see themselves as incompetent, passive, and with low self-esteem, and are viewed similarly by the parents. These perceptions tend to perpetuate the substance abuse.

The structure of addict families as described by clinicians show a high degree of enmeshment between the patient and some family members, excessive disengagement by others, as well as a reversal of the generational hierarchy (Madanes *et al.* 1980).

Despite the patient's disclaimer to the contrary, empirical studies have noted the frequent contact between patient and parents (Stanton *et al.* 1982). Attardo (1965) rated the level of symbiosis and separation-individuation in mothers of families with a drug-abusing member compared with schizophrenia and a control group. The mothers of drug-abusers scored highest on their need for symbiosis. It cannot be assumed that these relationships are pathogenic, but they may perpetuate the substance abuse.

Three patterns of sibling relationships have been described in the care of adolescents with drug abuse (Levine 1985).

1. The parents are disengaged from their children, who then turn to each other for support and become enmeshed, leading to co-abuse.

2. Generational boundaries are breached, an older sibling replacing an absent parent (usually father), thereby forming a parental relationship with the drug-abusing adolescent. Occasionally, it is the parentified child who becomes the drug-abuser.

3. Since the 'good' or 'successful' family roles or family 'niches' have been filled, the adolescent avoids competing for equal status as an achiever, and/or responds to a perceived family need for a scapegoat or 'failure' to contain the negative projections of the parents.

The adolescent peer group is widely regarded as a cogent influence in the development of drug-abuse. While the peer group most commonly introduces the adolescent to drugs (including illicit ones), the progression beyond experimental or social use to abuse is less a reflection of peer group pressure than of family dysfunction and individual psychopathology (Newcombe and Bentler 1989).

Environmental stress plays a role in the development of drug abuse (Osterweis *et al.* 1979). In migrant families, for example, the child who serves as 'gatekeeper' between the dominant culture and that of his family is vulnerable, especially if the latter's values and ways of life are markedly different from those of the wider culture. Death of a parent, usually father, during childhood is associated with increased risk, as is unresolved grief for a lost family member, or the recent death or loss by other means of an intimate friend (Coleman *et al.* 1986).

Family of procreation

Approximately one-third of married drug-abusing-patients have a spouse who is also a substance abuser (either alcohol or drugs). Even where the spouse does not abuse, over half the marriages are poor, with frequent separations and threat of divorce. Personality and interpersonal problems are evident in addition to the substance abuse. Motivation of patient or spouse for marital or family therapy is slight.

Patterns of drug abuse and family structure

A team of Italian psychiatrists has proposed a useful typology of heroin abuse which attempts to correlate the behavioural pattern, course, environmental stressors, and family structure (Cancrini *et al.* 1988). The resultant four types are:

1. Traumatic – the onset is sudden and rapidly dominates the person's life. The problem emerges in the context of an adolescent separation crisis or other traumatic event, such as loss through death or divorce. The person seeks numbness rather than pleasure, increases the dose rapidly and is

at risk for suicide. The underlying dynamic is one of self-punishment or incurring societal disapproval to atone for guilt induced by the trauma or separation crisis. The emphasis in management is individual psychotherapy combined with reconnecting the patient to his social network.

2. Developmental conflict – the family structure is typically enmeshed, particularly with over-involvement of the opposite-sex parent, and a peripheral role of the other parent. The patient is infantilized, inappropriately involved in the life of his parents, while his own life is filled with conflict or disappointment. The abuser's behaviour stabilizes the family. Individual psychotherapy is unlikely to be effective here; the family is the principal focus of treatment.

3. Transitional – drugs are used here to deal with deep personal suffering, often severe mood disorder. The compulsive drug-taking aims to numb feelings. The family's communication pattern is often unclear or oblique and revolves around the definitions of success and failure of the children. In addition to psychotropic medication, specific family therapy techniques such as paradoxical therapy may be required to challenge the family in a non-destructive way.

4. Sociopathic – this group represents the stereotype of the streetwise drug abuser. Family life is emotionally impoverished, chaotic, often violent, and rejecting. Antisocial behaviour begins early in life and precedes the abuse. Heroin is only one of many drugs used, with disregard for the physical damage they cause. Methadone maintenance therapy combined with participation in a therapeutic community may exert a cumulative beneficial effect if persevered with over an extended period.

TREATMENT AND PREVENTION

Stanton *et al.* (1982) reported the impressive superiority of results obtained by brief (10 sessions) structural-strategic family therapy compared with methadone maintenance and individual counselling in a controlled study. Twelve months after termination, patients in the family treatment groups were using significantly less illegal drugs and alcohol. Marked improvement in family interaction occurred in parallel, for example, increased involvement of fathers. However, the dropout rate was high in both treatment and control groups. Several well-designed studies of structural-strategic family therapy have attempted to remedy the methodological and conceptual problems encountered by the Stanton group (Szapocznik 1989). These confirm the efficacy of this model of family therapy in the treatment of adolescent drug addiction, and also show a differential rate of change of family members, with the substance abuser changing first, and others changing only after termination. Furthermore, the utility of individual combined with family therapy for highly 'dysfunctional' or 'resistant' families has been demonstrated.

Other studies combine individual, group, and family therapy (mainly structural-strategic). As with alcoholism, no single treatment method is universally effective, combined treatments showing the greatest promise (Piercy and Frankel 1989). However, many problems remain in determining the appropriate place of family therapy (Gleeson 1991; Reichelt and Christensen 1990), and such therapy will need adjustment to accommodate specific contextual and psychological factors (Kang *et al*. 1991; Nace *et al*. 1991).

The family may also play an important *preventative* role by protecting the adolescent from assuming psychosocial roles and tasks prematurely (for example, through parentification) and by discouraging the use of milder, socially acceptable substances. The latter may be combined with community education about the adverse effects of drugs. The family may obviously also bring its resources and problem-solving capacities to bear in the care of the adolescent who resorts to alcohol or drugs to deal with personal difficulties. Such preventative measures assume even greater importance in the context of community attempts to reduce the incidence of AIDS by discouraging needle-sharing among heroin users.

REFERENCES

Attardo, N. (1965). Psychodynamic factors in the mother-child relationship in adolescent drug addiction: a comparison of mothers of schizophrenics and mothers of normal adolescent sons. *Psychotherapy and Psychosomatics*, **13**, 249–55.

Bowen, M. (1974). Alcoholism viewed through family systems theory and family psychotherapy. *Annals of the New York Academy of Science*, **233**, 115–22.

Cancrini, L., Cingopani, S., Compagnoni, F. *et al*. (1988). Juvenile drug addiction: a typology of heroin addicts and their families. *Family Process*, **27**, 261–71.

Coleman, S.B., Kaplan, J.D., and Downing, R.W. (1986). Life-cycle and loss: the spiritual vacuum of heroin addiction. *Family Process*, **25**, 5–24.

Davis, S.K. (1990). Chemical dependency in women: a description of its effects and outcome on adequate parenting. *Journal of Substance Abuse Treatment*, **7**, 225–32.

Davis, T.S. and Hagood, L. (1979). In-home support for recovering alcoholic mothers and their families: the family rehabilitation and co-ordination project. *Journal of Studies on Alcohol*, **40**, 313–17.

Dinwiddie, S. H. and Cloninger, C.R. (1989). Family and adoption studies of alcoholism. In *Alcoholism: biomedical and genetic aspects*. (ed. H.W. Goedde and D.P.A. Agarwal). Pergamon, New York.

Galanter, M. (ed.) (1989). *Recent developments in alcoholism, Vol. 7: treatment research*. Plenum, New York.

Gleeson, A. (1991). Family therapy and substance abuse. *Australian and New Zealand Journal of Family Therapy*, **12**, 91–8.

Gliedman, L.H., Rosenthal, D., Frank, J.D., and Nash, H.T. (1956). Group therapy of alcoholics with concurrent group meetings with their wives. *Quarterly Journal of Studies on Alcohol*, **17**, 655–61.

Hasin D.S., Grant, B., and Endkott, J. (1990). The natural history of alcohol abuse: implications for definitions of alcohol abuse disorders. *American Journal of Psychiatry*, **147**, 1537–41.

Hesselbrock, V.M. (1986). Family history of psychopathology in alcoholics: A review and issues. In *Psychopathology and addictive disorders* (ed. H. Meyer). Guilford, New York and London.

Hunt, G.M. and Azrin, N.H., (1973). A community reinforcement approach to alcoholism. *Behaviour Research and Therapy*, **11**, 91–104.

Hurley, D.L. (1991). Women, alcohol and incest: An analytical review. *Journal of Studies on Alcohol*, **52**, 253–68.

Jackson, J.K. (1954). The adjustment of the family to the crisis of alcoholism. *Quarterly Journal of Studies on Alcohol*, **15**, 562–86.

Jacob, T. (1992). Family studies of alcoholism. *Journal of Family Psychology*, **5**, 319–38.

Jacob, T. and Krahn, G. (1988). Marital interaction of alcoholic couples. Comparison of depressed and non-distressed couples. *Journal of Consulting and Clinical Psychology*, **56**, 73–9.

Kang, Y.S., Kleinman, P.H., Woody, G.E. *et al.* (1991). Outcomes for cocaine abusers after once-a-week psychosocial therapy. *American Journal of Psychiatry*, **148**, 630–5.

Kaufman, E. (1984). *Power to change: family case studies in the treatment of alcoholism*. Gardner Press, New York.

Khatzian, E.J. (1985). The self-medication hypothesis of addictive disorders: Focus on heroin and cocaine dependence. *American Journal of Psychiatry*, **142**, 1259–64.

Levine, B.L. (1985). Adolescent substance abuse: toward an integration of family systems and individual adaptation theories. *American Journal of Family Therapy*, **13**, 3–16.

Madanes, C., Dukes, J., and Harbin, H., (1980). Family ties of heroin addicts. *Archives of General Psychiatry*, **3**, 889–94.

McCance-Katz, E.F. (1991). The consequences of maternal substance abuse for the child exposed in utero. *Psychosomatics*, **32**, 268–74.

de Maio, R., (1989). Integrating traditional alcoholic treatment programme and family-systems therapy. *Family Systems Medicine* **7**, 274–91.

Moos, R., Finney, J., and Gamble, W., (1982). The process of recovery from alcoholism: II. Comparing spouses of alcoholic patients and matched community controls. *Journal of Studies on Alcohol*, **43**, 888–909.

Nace, E.P., Davis, C.W., and Gaspari, J.P. (1991). Axis II co-morbidity in substance abusers. *American Journal of Psychiatry*, **148**, 118–20.

Newcomb, M.D. and Bentler, P.M. (1989). *Consequences of adolescent drug abuse*. Sage, Newbury Park, CA.

O'Farrell, T.J. (1992). Families and alcohol problems: an overview of treatment research. *Journal of Family Psychology*, **5**, 339–59.

Orford, J., Oppenheimer, E., Egert, S. *et al.* (1976). The cohesiveness of alcoholism-complicated marriages and its influence on treatment outcome. *British Journal of Psychiatry*, **128**, 318–39.

Orford, J. (1985). Alcohol problems and the family. In *Approaches to addiction*, (ed. J. Lishman and G. Horobin). Kogan Page, London.

Orford, J. (1990). Alcohol and the family: an international review of the literature with implications for research in practice. In *Research advances in alcohol and drug problems, Vol. 10* (ed. L.T. Kozlowski). Plenum, New York.

Osterweis, M., Bush, P.J., and Zuckerman, A.E. (1979). Family context as a predictor of individual medicine use. *Social Science and Medicine*, **13**, 287–91.

Piercy, F.B. and Frankel, B.R. (1989). The evolution of an integrative family therapy for substance-abusing adolescents: toward the mutual enhancement of research and practice. *Journal of Family Psychology*, **3**, 5–25.

Reichelt, S. and Christensen, B. (1990). Reflections during a study on family therapy with drug addicts. *Family Process*, **29**, 273–87.

Sandmaier, M. (1980). *The invisible alcoholics: women and alcohol abuse in America*. McGraw Hill, New York.

Stanton, M.D. (1979). Drugs and the family. *Marriage and Family Review*, **2**, 1–10.

Stanton, M.D., Thomas, C. *et al.* (1982). *The Family therapy of drug abuse and addiction*, Guilford Press, New York.

Steinglass, P. (1979). Family therapy with alcoholics: a review. In *family therapy of drug and alcohol abuse* (ed. E. Kaufman and P. Kaufmann). Gardner Press, New York.

Steinglass, P. (1980). A life history model of the alcoholic family. *Family Process*, **19**, 211–26.

Szapocznik, J. I. and Kurtines, W.M. (ed.) (1989). *Breakthroughs in the family therapy with substance abusing and problem youth*. Springer, New York.

Thomas, E.J. and Santa, C.A. (1982). Unilateral family therapy for alcohol abuse: a working conception. *American Journal of Family Therapy*, **10**, 49–58.

Uhl, G.R., Persico, A.M., and Smith, S.S. (1992). Current excitement with D_2 dopamine receptor gene alleles in substance abuse. *Archives of General Psychiatry*, **49**, 157–61.

Wolin, S.J., Bennett, L.A., Noonan, D.L., and Teitelbaum, M.A. (1980). Disruptive family rituals: a factor in the intergenerational transmission of alcoholism. *Quarterly Journal of Studies on Alcohol*, **41**, 199–214.

Ziegler–Driscoll, G. (1979). The similarities in families of drug dependants and alcoholics. In *Family therapy of drug and alcohol abuse*. (ed. E. Kaufman and P. Kaufmann). Gardner Press, New York.

Zucker, R.A. and Lisansky–Gomberg, E.S. (1986). Etiology of alcoholism reconsidered. *American Psychologist*, **41**, 783–93.

10

Anxiety disorders

In planning this chapter, a number of issues specific to anxiety disorders became apparent to us. First, research on marital and family therapy has been done only in relation to agoraphobia and obsessive-compulsive disorder (OCD), so we have focused almost exclusively on these. The work is largely concerned with couples therapy, reflecting the fact that most patients with agoraphobia and OCD first present for treatment in adulthood. Second, agoraphobia and OCD are disorders so different from each other that they require separate consideration. Third, although there is a modest literature on the familial and genetic aspects of the two disorders, it is not methodologically robust, and there are few consistent findings. This is primarily because a consensus on diagnosis and classification of anxiety disorders has only emerged since the 1980s.

We have therefore decided to omit discussion of familial and genetic aspects of agoraphobia and OCD, allowing more space for a consideration of the process and outcome of marital and family therapy. Knowledge has grown rapidly, a growth fuelled by the success of behaviour therapy in treating those formerly relatively refractory conditions. Paradoxically, rapid and substantial improvement after behaviour therapy revealed negative repercussions on a proportion of spouses which sometimes impeded patients' further progress (Hafner 1977). Observations of this kind created interest in the marital context of agoraphobia and OCD, and it is clear that attention to this is often essential for optimal therapy.

PANIC DISORDER

In psychiatric practice, it is rare for panic disorder to present in an uncomplicated fashion. For example, a detailed study of an unbiased sample of 90 patients who met DSM-III criteria for panic disorder revealed that only 6 of the 90 had that disorder alone (Argyle and Roth 1989). Fifty-six also had agoraphobia, and a further 20 suffered limited avoidance of an agoraphobic nature. Thirty-seven patients had social phobia, of whom 26 also had agoraphobia. Major depressive episode was concurrently diagnosed in 35 patients, and clinically significant depression in a further 16. Only two patients failed to meet the criteria for generalized anxiety disorder. These findings are important, because they reflect not only a consensus on co-morbidity, but also the clinical reality of routine psychiatric practice.

Individual cognitive or cognitive-behaviour therapy is a highly effective treatment for uncomplicated panic disorder, although the addition of tricyclic medication is often desirable when the panic attacks are severe. Because of the excellent results with individual therapy, there has been no impetus toward the development of marital or family therapy for uncomplicated panic disorder; it seems likely that a family approach is unnecessary in most cases. However, when panic disorder is associated with agoraphobia, or with another disorder, a family approach is more likely to be indicated.

PANIC DISORDER WITH AGORAPHOBIA

The relationship between panic disorder and agoraphobia remains blurred. Clinically, most patients presenting with agoraphobia give a clear history of panic attacks, and it is assumed that the agoraphobia develops primarily as a means of avoiding these attacks. However, community surveys have found that panic attacks and agoraphobia are associated in only a minority of cases, casting doubt upon the aetiological role of panic attacks. Moreover, the Epidemiologic Catchment Area Survey in the USA found a six-month prevalence rate for agoraphobia of 3.7 per cent, whereas that for panic disorder was only 1.7 per cent, with a co-morbidity rate of 0.7 per cent (Regier *et al.* 1988). Thus, aetiological factors other than panic attacks are clearly operating in community cases. Agoraphobia is three to four times more common in women than in men, whereas panic disorder is less than twice as common. If panic disorder is often a precursor to agoraphobia, then this occurs much more frequently in women.

DSM-III-R creates two categories of agoraphobia, one with and one without panic disorder. However, most research into agoraphobia was carried out before this distinction was made, and in describing the relevant studies we use the term agoraphobia alone.

RESEARCH ON THE MARITAL CONTEXT OF AGORAPHOBIA

Almost all studies of agoraphobia are based primarily or exclusively on married women, who make up 50–60 per cent of the clinical population. Although married women with agoraphobia have been widely studied, there is much controversy about the relationship between marital dynamics and the condition, a controversy originating in the different perspectives of clinician and researcher. Clinicians who base their reports on practice experience almost always suggest an important role for marital dynamics in the disorder's development (Holmes 1982). Investigators who design experimental studies but have little direct involvement in treating agoraphobia, regard marital dynamics as irrelevant (Arrindell 1987). The controversy is unfortunate because it has confused clinicians about the role of marital and family therapy in treatment. It is important to examine its basis in detail, so that rational judgements can be made about indications for family therapy.

Case selection bias and false generalization

The key to understanding the controversy lies in case selection. Clinicians attempt to treat most of those patients referred to them. They are willing to take on severe and complex cases, including, for example, agoraphobia complicated by major depression, personality disorder, and other problems. Research on co-morbidity suggests that over 60 per cent of agoraphobic patients presenting for specialist treatment have at least one additional psychiatric condition, of which major depression, dysthymia, social phobia, and personality disorder are the most common. A study by Chambless *et al.* (1992) found that no less than 90 per cent of 165 agoraphobic patients attending an anxiety treatment centre had personality disorders, of which the most common were avoidant and dependent. Co-morbidity significantly reduces the chances of a good treatment outcome, both directly and by association with marital dysharmony. When therapists comment on their work, it obviously reflects their clinical practice, within which negative interactions between agoraphobia, personality disorder, and marital dysharmony are common.

Those who systematically research agoraphobia work with patients who are different from those seen by clinicians. Almost all studies have utilized rigid *exclusion* criteria, whereby agoraphobic subjects are rejected if they exhibit significant depression, social phobia, personality disorder, drug abuse, or indeed any other significant symptoms or problems (Zitrin *et al.* 1983). From a scientific perspective, it is logical to research a sample that is as 'pure' as possible. However, in doing this, so many cases are excluded or exclude themselves that the sample becomes unrepresentative of the clinical population. Few researchers publish details of the number of patients excluded. However, Jannoun *et al.* (1980) reported that only 28 of 53 patients (53 per cent) referred to an experimental treatment programme for agoraphobia actually started therapy. This level of rejection by researchers or self-exclusion by patients probably occurs prior to most experimental studies. In addition, an average 25 per cent of patients drop out of such studies (Mavissakalian and Michelson 1982), so that those completing treatment are a minority, selected in rarely specified or understood ways. Nonetheless, findings are almost always generalized to the entire population of agoraphobics. These 'pure' cases commonly respond well to the treatments involved, usually various combinations of drugs and (cognitive) behaviour therapy. As a result, these individual approaches have become overvalued. In reality, they are rarely sufficient in the complex cases that make up a considerable proportion of the clinician's case load.

This bias is even greater in experimental studies that involve the spouse. Any substantial degree of marital conflict reduces the spouse's willingness to participate (Hafner 1989). Thus, studies comparing couple and individual treatment include mainly uncomplicated cases with harmonious marriages. Such samples would not be expected to benefit from a marital approach (see Chapter 5), but negative findings in this regard are invariably generalized, leading to the false idea that couple therapy for agoraphobia is unhelpful.

The work of Cobb *et al*. (1984) is a good example of this. They compared individual and couple therapy in 18 married agoraphobics. Finding no advantage for a couple approach, they concluded that 'agoraphobia would appear to be primarily a problem arising from the individual'. However, the patients' mean pre-treatment score on a measure of marital conflict resembled the normal population, indicating a reasonable level of marital satisfaction; it is, therefore, not surprising that couple therapy was of little extra benefit.

Another example of false generalization is the influential research of Arrindell and Emmelkamp, and their coworkers. Arrindell (1987) has examined the marital adjustment of agoraphobic women and its effect on the outcome of behaviour therapy. He concludes that marital (and personal) adjustment is normal, and that attention to the marriage is unnecessary. However, in his study of marital conflict in agoraphobia, only *one* couple out of 25 had a poor marriage: yet Arrindell generalizes his findings to all agoraphobic women, including those with marital conflict.

He then dismisses the extensive research that contradicts his findings on the basis that 'The experimenter as therapist is a serious methodological problem in marital treatment research; the emotional investment and enthusiasm of the author/therapist may not only . . . spuriously inflate the success of treatment under evaluation, but may also lead to selective sampling and description of patients'.

Emmelkamp *et al*. (1992) in comparing spouse-aided and individual therapy in 60 agoraphobic patients were unable to find enough maritally dissatisfied patients for separate statistical analysis. Therapy comprised four sessions and involved discussion of patient progress at self-exposure, guided by a self-help manual. In the spouse-aided condition, spouses were included in discussions, but a focus on any relationship problems was postponed until after the study's completion. Not surprisingly, treatment effects were weak, the authors themselves pointing out that improvement in the agoraphobia was 25–30 per cent less than that obtained in other comparable studies of exposure-*in-vivo*. No benefits accrued from the spouse's involvement, which is to be expected from such a treatment in a maritally satisfied population. In recommending individual treatment, Emmelkamp *et al.* generalize their findings to the total agoraphobic population; at no stage do they consider that their sample might not be representative.

Because this sort of work is based on relatively large numbers and appears methodologically robust, it tends to be far more influential than smaller scale or in-depth studies. Unless it is recognized that research samples are generally unrepresentative, myths about treatment effectiveness will arise.

In spite of these difficulties, several empirical studies suggest the utility of a couple approach. The following is the best designed of these.

The study by Barlow's group

The patients, all married women, suffered from moderate to severe agoraphobia and were drawn mainly from an anxiety disorder clinic (Cerney *et al.* 1987). Treatment was based on group discussions and exposure-*in-vivo* homework assignments over a 12-week period. In the spouse-aided group ($N = 28$), husbands attended all sessions and were actively involved. In the individually treated group, husbands were excluded. Patients in the spouse-aided group did significantly better than those in the individual group, who, at the 12-month follow-up, had deteriorated slightly. At a two-year follow-up, this superiority was maintained or increased. The authors concludes that 'These data would seem to confirm the importance of attending to the interpersonal context of behaviour change, in this case, by including the spouse . . .'.

The same team studied communication patterns between marital partners during exposure-*in-vivo* (Craske *et al.* 1989). They found that frequent communication was predictive of a good outcome. This compliments the methodologically rigorous work of Arnow *et al.* (1985), who compared couple communication training and couple relaxation training as a supplement to partner-assisted exposure-*in-vivo*. The response to communication training was significantly better, a superiority maintained at eight-month follow-up. These studies suggest that involving the spouse in treatment is likely to be of benefit when it results in improved communication.

Outstanding research issues

Little is known about indications and contra-indications for a couple approach to treating agoraphobia. Although patients with significant marital difficulties seem likely to benefit most, the relationship between treatment outcome and marital adjustment is unclear. Patients with bad marriages appear to respond poorly to individual treatment, with the reverse true of couple therapy. Some workers have found no relationship between marital adjustment and outcome. This is an important area for elucidation, because there is an urgent need to improve the effectiveness of existing treatments. Only 20–30 per cent of patients finish exposure-based treatments substantially free of agoraphobia (Jacobson *et al.* 1988). When used as the sole treatment, the effectiveness of medication is little better, especially when compliance problems are taken into account.

Another controversial area concerns the health of the patient's spouse. As we noted in the introduction, early researchers found rapid improvement in married women with agoraphobia after exposure-*in-vivo* to be associated with deterioration in their husbands' health. These health problems not infrequently impeded or reversed the patients' response to treatment. This has not been replicated since the 1970s except for Oatley and Hodgson's (1987) important study, and it is a problem that occurs mainly with highly effective treatments such as intensive group exposure-*in-vivo*. During these, the patient's improvement may be prompt and substantial. The marital system may not have time to adapt, increasing the chance of a negative effect on the husband. This is particularly relevant when the husband has his own pre-existing psychosocial problems; husbands most at risk are those married to women with severe, long-standing agoraphobia complicated by depression, personality disorder, and marital dissatisfaction. Including the husband in therapy reduces the likelihood of negative repercussions.

PRACTICAL APPROACHES TO FAMILY THERAPY FOR AGORAPHOBIA

The focus of what follows is on married women, but much of it applies to the family treatment of young unmarried patients living with their parents. There is little work concerning married men with agoraphobia, but some evidence that they have special needs; we comment on this later.

A reasonable consensus exists that exposure-*in-vivo* is central in recovery from agoraphobia. An unresolved issue concerns how to ensure that optimal exposure occurs. In less severe cases, it seems that detailed therapist instruction and firm encouragement to enter feared situations is sufficient. In more severe cases, the patient needs to be accompanied into feared situations, at least initially. The accompanying person may be the therapist, another trained mental health professional, a trained volunteer (perhaps an ex-agoraphobic), a fellow agoraphobic (or a small group of agoraphobics), a member of a support group, or a friend or relative. Husbands appear less effective in this role than friends or members of a support group.

In at least a third of cases, agoraphobia is a severe, disabling condition that resists standard treatment. As we have seen, this is especially true when it is complicated by depression, social phobia, personality disorder, hypochondriasis, and marital conflict, which tend to co-exist. Research papers on agoraphobia report a delay of some 6–10 years between onset and attendance at a specialized treatment clinic. Most attendees will have delayed consultation for as long as possible, often because they do not wish confirmation of their fear of insanity. Others will have attended a range of mental health professionals, often repeatedly. A clinician seeing a married woman with severe agoraphobia for the first time is likely to encounter problems establishing a treatment alliance, and even more so with her husband. These problems may be summarized as follows:

1. Previous attempts at therapy will almost always have been individual rather than conjoint, and may have consolidated the patient's view that she is suffering from a medical condition over which she has little control. During individual therapy, the patient's husband often becomes alienated, either because treatment appears ineffective (or seems to exacerbate the condition), or because he is made to feel blameworthy. It is common for the patient to subtly misrepresent her therapist's comments. For example, brief discussion of divorce may be relayed to the husband as 'My psychiatrist says that I should divorce you'. As well, the patient may emphasize how understanding her psychiatrist is, with the implication that he meets her needs better than her husband does. Such statements may enrage the husband and make him envious toward the therapist: they can also make him feel hopeless about understanding his wife's disorder, frustrated about helping her manage it, and inadequate as a person.

2. The patient may be reluctant to involve her husband in therapy. She may not wish to lose her own exclusive therapist, and may fear also that any previous misrepresentation of therapy will be exposed. Most importantly, she fears that the power of her husband over her will be increased, a power already greater than usual because of her agoraphobia-related dependency. Indeed, the husband may perceive conjoint therapy as a chance to impose his views or to become more powerful in other ways. He will also often consciously or unconsciously seek to undermine the

therapist's competence and authority, thereby reducing any feelings of envy or resentment.

3. We outlined in Chapter 5 why using the term marital therapy often alienates couples who have grappled for several years with chronic, disabling depression in one partner. This is especially true of the spouse, and the husbands of women with severe agoraphobia may be even more difficult to engage in marital therapy than the husbands of severely depressed women. For example, Forrest (1969) was able to engage only one of 18 men. Clinical references to this difficulty still occur frequently, although other therapists claim that it is not a problem. Presumably, they are dealing with patients in good marriages and with uncomplicated agoraphobia.

It is easy to avoid the term marital therapy when discussing the husband's involvement. It can be stressed that the husband's task is to facilitate the patient's exposure-*in-vivo*, and to help her manage related difficulties. To avoid being duplicitous, it is vital to point out to the husband that his previous attempts to help his wife may inadvertently have had the opposite effect. Because of this, it is necessary to look at his coping and communication skills, and to seek ways of improving them. This, it must be explained, will probably involve self-disclosure on his part, and perhaps exploration of his own adjustment.

Attempts at couple therapy which grapple with the above issues have a reasonable chance of success. One approach, termed spouse-aided therapy, has been described elsewhere (Hafner 1988*a*,*b*). A randomized comparison of spouse-aided and individual therapy in severe, persisting psychiatric disorders included eight patients with agoraphobia. Results showed that spouse-aided therapy was superior to individual therapy for both patients and spouses (Hafner *et al.* 1983). What follows is an outline of the approach.

Spouse-aided therapy for agoraphobia

The role of medication

The role of medication is clarified at the outset. Ideally, patients should withdraw from benzodiazepines before starting systematic exposure, since there is evidence of state-dependence with these drugs that inhibits generalization of learning effects. The task of withdrawal may be so challenging that it becomes a major dimension of therapy. Patients who do not wish to withdraw can, of course, proceed with exposure, but the likelihood of successful therapy diminishes. Patients who have shown partial improvement with tricyclic antidepressants should probably remain on them during exposure-*in-vivo*. Most behaviour therapists however feel that best results with exposure-*in-vivo* occur when the patient is entirely or substantially free of medication.

Establishing treatment goals

Setting realistic treatment goals is fundamental, with both partners actively involved. Often, before goals can be set, a session is devoted to educating the couple about the realities of agoraphobia. Explaining the power of negative reinforcement is important. Most patients have repeatedly attempted self-exposure. Often, these attempts are terminated by a sudden escape from the feared situation, precipitated by acute anxiety or panic. Anxiety levels fall rapidly and this rapid decrease negatively reinforces the escape/avoidance pattern. This has the effect of exacerbating the condition, especially the anticipatory anxiety, or 'fear of fear', which is central in most cases. In general, returning home reduces anxiety rapidly. This means that negative reinforcement of the home as a place of safety is almost inevitable, with a corresponding increase in fear of the outside world. These self-perpetuating aspects are crucial, and it is essential for the couple to understand them if they are to work together effectively.

For the spouse, establishing the genuine nature of the patient's symptoms is often the most important aspect of the educational phase; the symptoms may fluctuate markedly from day to day or week to week, which makes it difficult for spouses (and patients) to regard the condition as 'genuine', much less persistent. With the spouse's understanding, he can take part in setting treatment goals. These are concerned with developing optimal strategies for exposure-*in-vivo*. Explaining learning theory is crucial, including how to construct a hierarchy of fears and moving up it *only* when lesser fears have been substantially resolved. Treatment manuals devised by Mathews *et al.* (1981) are helpful in this regard.

Achieving an interpersonal focus

Inevitably, conflict arises during the couple's attempts at working toward treatment goals, which as a result may not be achieved. These conflicts allow the therapist to focus on the couple's relationship. Since a therapeutic alliance will have been established, it is possible to clarify with either or both partners their contribution to marital difficulties. Their origins are enormously varied. Resolving them is often the crux of successful therapy. It is common for couple, patient or spouse to blame the therapist for failure to achieve goals. A reminder of the couple's own role in determining these enables them to acknowledge their own contribution to such failure. This 'challenge' is most effective when linked to a clear expression of the therapist's own limitations in the absence of a working alliance. Problems commonly regarded as reflecting 'Oedipal' and 'transference' issues are likely to crystallize during this phase. It is rarely appropriate to interpret these to couples but awareness of them by the therapist is often essential for optimal intervention.

Occasionally, the 'challenge' reveals personal problems in patient or spouse that make it impossible for goals to be achieved. When this happens, it is important to avoid blaming either partner for the impasse. Instead, other options should be discussed, and implemented if possible. In such circumstances, the couple may agree to leave matters unresolved, accepting limited improvement.

Although worthwhile benefits occur in most cases, this often reveals interpersonal problems that were previously submerged. Some couples choose at this point to embark on marital therapy. Not infrequently, the spouse decides to enter individual therapy. Occasionally, couples choose to initiate separation or divorce. The therapist has an important role in facilitating the next phase of therapy, which will often involve other therapists.

Married men with agoraphobia

Married men with agoraphobia are rare in clinical practice. In most studies, 70–80 per cent of patients are women, with married men representing only 5–10 per cent of the sample.

Sex role stereotyping is probably important in the genesis of severe agoraphobia in married men. The male stereotype discourages open expression of fearfulness. This means that husbands hesitate to disclose the full extent of their pre-agoraphobic symptoms. However, when severe panic attacks develop, they cannot easily disguise their physiological and behavioural manifestations, to which wives usually respond more sympathetically than to overt expression of anxiety or fear. As a result, wives may selectively reinforce the physiological concomitants of panic attacks. These include palpitations, choking sensations, shortness of breath, dizziness, unsteadiness, shaking, and sweating. Often, the symptoms become the focus of medical investigations aimed at excluding serious physical illness. This focus may provide further reinforcement. Frequently, hypochondriasis emerges as a major aspect of the picture, and this may delay or prevent the diagnosis of panic disorder. As long as the condition remains undiagnosed, the panic attacks evoke acute fear in both patient and wife. The situation may be further complicated by the use of alcohol as self-medication, a process encouraged by the male sex role stereotype.

Either as part of these developments or in relation to panic attacks *per se*, the agoraphobic man becomes dependent on his wife. Ostensibly, this is connected to a fear of dying or of totally losing control during a panic attack. In extreme cases, husbands need either to be with their wives constantly or to know their exact whereabouts at all times, so that they may be contacted if a panic attack looms or eventuates. At a dynamic level, this recapitulates the intense, ambivalent dependency that these men experienced in their relationship with their mothers. However, as adults they have an image of themselves that requires them to deny dependency. Unconsciously, they wish to be both powerful and protective and dependent and submissive. These incompatible expectations contribute to marital conflict, and this in turn helps to perpetuate the panic attacks. However, once the agoraphobia develops, this conflict is resolved or ameliorated: wives become protective in relation to their husband's panic attacks and fears, but, at the same time, husbands control their wives by insisting they devote themselves to protecting them from panic attacks and their consequences.

Such couples insist that the panic be the exclusive focus of therapy. However, adhering to the principles of spouse-aided therapy facilitates an interpersonal perspective. This creates an opportunity to address dynamic issues that underpin pathogenic marital interaction. The process is facilitated by dealing explicitly with sex role stereotypes and the ways in which these have contributed to symptom development.

OBSESSIVE-COMPULSIVE DISORDER

The Epidemiologic Catchment Area survey found a six-month prevalence rate for obsessive-compulsive disorder (OCD) of about 1.5 per cent (Karno *et al.* 1988). This rate was 25–60 times higher than estimates based on studies of clinical populations. Reasons for under-representation of OCD in clinical practice are unclear. However, patients are reluctant to reveal the nature and extent of their symptoms, and this is an obstacle to seeking treatment. Support and self-help groups have gained momentum in the 1980s, linked with increased public awareness. Consequently, an increasing proportion of OCD sufferers do seek treatment.

The effectiveness of behaviour therapy based on exposure-*in-vivo* and response prevention has been repeatedly demonstrated. Improvement rates of 60–70 per cent are usual, and these are generally maintained. However, many patients find behaviour therapy unacceptable or drop out prematurely so that it is helpful in only about 50 per cent of cases (Perse 1988). Similar constraints apply to treatment with clomipramine which has the additional problem of a high relapse rate on withdrawal (Pato *et al.* 1988). Thus, there is a need for treatment which is more effective.

The marital and family context of OCD

There is substantial marital and family disharmony in OCD (Emmelkamp *et al*. 1990; Tynes *et al*. 1990). Although family conflict revolves around symptoms, causal relationships have not been established. Clinically, it is usual to find a circular pattern between symptoms and family conflict.

The benefit of involving family members in the treatment of severe OCD is established (Hafner *et al*. 1981). However, most studies concern treatment of adolescents, and reports on family treatment in adult cases are rare (Hafner 1982; Tynes *et al*. 1990). The results of empirical studies have been mixed. Two early studies suggested that involving the spouse in a behavioural approach enhanced outcome. However, other work by Emmelkamp *et al*. (1990) failed to confirm their own previous results suggesting the benefits of a conjoint approach: there were no benefits from involving the spouse, either on symptoms or marital adjustment. This study was large, with 50 married out-patients randomly assigned to either spouse-aided or individual therapy; it is therefore likely to exert a significant influence on clinical practice.

Emmelkamp *et al*. involved spouses on strictly behavioural lines: 'The task of the partner was to encourage the patient and to have him confront the distressing stimuli until the patient got used to them. In addition, the partner was instructed to withhold reassurance if the patient asked for it'. This approach does not permit a focus on marital dynamics and their relationship to symptom development and maintenance. From a dynamic perspective, it seems likely to reinforce unhelpful marital interaction, probably reducing the chance of symptomatic improvement. For such reasons, experienced clinicians often object to involving the spouse as co-therapist. However, the risks of such an approach can be reduced or eliminated if it includes an exploration of marital dynamics (Hafner *et al*. 1983).

There has been another useful comparison of family and individual treatment. Mehta (1990), studying 30 patients, found that involvement of a family member in a behavioural approach yielded superior results on obsessive-compulsive symptoms and family adjustment. The treatment was more intensive and prolonged than that of Emmelkamp *et al*. (24 sessions over 12 weeks *versus* 8 sessions over 5 weeks). It appears to have been much more effective. An important difference between the two approaches concerns instructions given to relatives who, in the Mehta study, were '. . . instructed to be supportive when the patient was depressed and to allay the patient's anxieties'. This contrasts with the more confrontational approach of Emmelkamp. Together with the greater number of sessions, it may have allowed for changes in marital dynamics that could not occur in Emmelkamp's study.

Practical approaches to the family treatment of OCD

In severe OCD, the patient's rituals often take up a large part of each day, and are highly disruptive to family life. The obsessions are usually very distressing. In nearly all cases, both patient and family insist that symptoms be the exclusive focus of treatment. Often, previous individually oriented therapy has consolidated the belief that the patient is suffering from a biologically determined illness in relation to which family interaction is of little relevance. Therapy which focuses directly on marital or family relationships is likely to be rejected; even if it is not, during therapy it is usually difficult to shift the focus from symptoms. The spouse or parents are more likely to be engaged by an invitation to become involved as co-therapists, working directly on the symptoms. Usually, the patient is comfortable with this approach, although he must believe that the therapist will protect him from any hostility that members direct toward him. Once the family is engaged, the following themes may emerge.

'Compulsory marriage'

Many spouses state that they have thought of leaving the marriage; overt threats to do so may be part of their attempt to 'manage' symptoms and related problems. They usually explain that they have remained mainly out of sense of duty to help the patient and in the belief that the patient would be unable to survive without them, a belief which the patient shares. In such circumstances, the latter fears the consequences of symptomatic improvement, believing that the spouse will leave if not bound by an obligation to help her survive the daily struggle. Unless this issue is resolved, improvement is unlikely.

The spouse's denial of personal problems

Spouses inevitably bring their own problems into the marriage, and subsequently acquire new ones. Often, these problems are a cause of marital conflict before the OCD symptoms become prominent. Spouses may be agoraphobic or have social phobia; they may have unresolved bereavement reactions, sexual disorders, or troublesome obsessive-compulsive symptoms themselves. As the patient's OCD develops, it becomes the main focus of marital dissatisfaction, and the spouse is protected from examining his own problems. Projection, denial, and displacement are defensive manoeuvres commonly used here. Anger, guilt, grief, and depression can be displaced onto the patient's symptoms or directly onto the patient. Where a 'compulsory marriage' exists, the spouse is able to treat the patient in a way that might otherwise cause her to leave the marriage. Negative interactions surrounding symptoms become an entrenched aspect of family life, but they are usually denied by both partners, especially when physical violence is involved. When spouse-aided therapy reaches the stage of exposure-*in-vivo*, response prevention, or other behavioural techniques, increased negative interaction is inevitable. However, by this time the patient should be confident that the therapist will ameliorate spousal aggression or at least not add to it; and the spouse should have realized that the therapist will not blame him for his behaviour toward the patient. It then becomes possible to discuss negative interactions and to identify ways of ameliorating them. If this is not done, unhelpful interactions are apt to continue unabated. This impedes progress in therapy, or may reverse initial improvement.

SOCIAL PHOBIA, SIMPLE PHOBIA, AND POST-TRAUMATIC STRESS DISORDER

There is no empirical evidence favouring a couple approach to any of the remaining anxiety disorder categories in DSM-III-R. Clinically, however, it may be helpful to involve spouses or other family members in the treatment of social phobia, simple phobia, generalized anxiety disorder, and post-traumatic stress disorder. Decisions about whether or not to seek their involvement are made on a commonsense basis. In general, the more severe and complex the disorder, the more likely is relatives' involvement to be worthwhile.

REFERENCES

Argyle, N. and Roth, M. (1989). The phenomenological study of 90 patients with panic disorder, Part II. *Psychiatric Developments*, **3**, 187–209.

Arnow, B.A., Taylor, C.B., Agras, W.S. *et al.* (1985). Enhancing agoraphobia treatment outcome by changing couple communication. *Behaviour Therapy*, **16**, 452–67.

Arrindell, W.A. (1987). *Marital conflict and agoraphobia: fact or fantasy?* Eburon, Delft.

Cerney, J.A., Barlow, D.H., Craske, M.G. *et al.* (1987). Couples treatment of agoraphobia: a two-year follow-up. *Behaviour Therapy*, **18**, 401–15.

Chambless, R.L., Renneberg, B., and Goldstein, A. (1992). MCMI-diagnosed personality disorders among agoraphobic patients: prevalence and relationship to severity and treatment outcome. *Journal of Anxiety Disorders*, **6**, 193–211.

Cobb, J.P., Mathews, A.M., Childs-Clarke, A. *et al.* (1984). The spouse as co-therapist in the treatment of agoraphobia. *British Journal of Psychiatry*, **144**, 282–7.

Craske, M.G., Burton, T., and Barlow, D.H. (1989). Relationships among measures of communication, marital satisfaction and exposure during couples treatment of agoraphobia. *Behaviour Research and Therapy*, **27**, 131–40.

Emmelkamp, P.M.G., de Haan, E., and Hoogduin, C.A.L. (1990). Marital adjustment and obsessive-compulsive disorder. *British Journal of Psychiatry*, **156**, 55–60.

Emmelkamp, P.M.G., van Dyck, R., Bitter, M. *et al*. (1992). Spouse-aided therapy with agoraphobics. *British Journal of Psychiatry*, **160**, 51–6.

Forrest, A.D. (1969). Manifestations of 'hysteria': phobic patients and hospital recidivists. *British Journal of Medical Psychology*, **42**, 263–70.

Hafner, R.J. (1977). The husbands of agoraphobic women and their influence on treatment outcome. *British Journal of Psychiatry*, **129**, 378–83.

Hafner, R.J. (1982). Marital interaction in persisting obsessive-compulsive disorders. *Australian and New Zealand Journal of Psychiatry*, **16**, 171–8.

Hafner, R.J. (1988*a*). Marital and family therapy. In *Handbook of anxiety disorders* (ed. C.G. Last and M. Hersen). Pergamon, New York.

Hafner, R.J. (1988*b*). Anxiety disorders. In *Handbook of behavioural family therapy* (ed. I.R.H. Falloon). Guilford, New York.

Hafner, R.J. (1989). Sex role stereotyping in women with agoraphobia and their husbands. *Sex Roles*, **20**, 705–11.

Hafner, R.J., Gilchrist, P., Bowling, J. *et al*. (1981). The treatment of obsessional neurosis in a family setting. *Australian and New Zealand Journal of Psychiatry*, **14**, 145–51.

Hafner, R.J., Badenoch, A., Fisher, J. *et al*. (1983). Spouse-aided versus individual therapy in persisting psychiatric disorders: a systematic evaluation. *Family Process*, **22**, 385–99.

Holmes, J. (1982). Phobia and counterphobia: family aspects of agoraphobia. *Journal of Family Therapy*, **4**, 133–52.

Jacobson, N.S., Wilson, L., and Tupper, C. (1988). The clinical significance of treatment gains resulting from exposure-based interventions for agoraphobia: a reanalysis of outcome data. *Behaviour Therapy*, **19**, 539–54.

Jannoun, L., Munby, M., Catalan, J. *et al*. (1980). A home-based treatment programme for agoraphobia: replication and controlled evaluation. *Behavioural Therapy*, **11**, 294–305.

Karno, M., Golding, J.M., Sorenson, S. *et al*. (1988). The epidemiology of obsessive-compulsive disorder in five US communities. *Archives of General Psychiatry*, **45**, 1094–9.

Mavissakalian, M. and Michelson, L. (1982). Agoraphobia: behavioural and pharmacological treatments, preliminary outcome, and process findings. *Psychopharmacology Bulletin*, **18**, 91–103.

Mathews, A.M., Gelder, M.G., and Johnston, D.W. (1981). *Agoraphobia: nature and treatment*. Tavistock, London.

Mehta, M. (1990). A comparative study of family based and patient-based behavioural management in obsessive-compulsive disorder. *British Journal of Psychiatry*, **157**, 133–5.

Oatley, K. and Hodgson, D. (1987). Influence of husbands on the outcome of agoraphobic wives' therapy. *British Journal of Psychiatry*, **150**, 380–6.

Pato, M.T., Zohar-Kadouch, R., Zohar, J. *et al*. (1988). Return of symptoms after discontinuation of clomipramine in patients with obsessive-compulsive disorder. *American Journal of Psychiatry*, **145**, 1521–5.

Perse, T. (1988). Obsessive-compulsive disorder: a treatment review. *Journal of Clinical Psychiatry*, **49**, 48–55.

Regier, D.A., Boyd, J.H., Burke, J. *et al.* (1988). One-month prevalence of mental disorders in the United States. *Archives of General Psychiatry*, **45**, 977–86.

Tynes, L.L., Salins, C., and Winstead, D.K. (1990). Obsessive-compulsive patients: familial frustration and criticism. *Journal of the Louisiana State Medical Society*, **142**, 28–9.

Zitrin, C.M., Klein, D.F., Woerner, M.G. *et al.* (1983). Treatment of phobias: 1. comparison of imipramine hydrochloride and placebo. *Archives of General Psychiatry*, **40**, 125–37.

11

Personality disorders

There are two difficulties in writing about the family and personality disorders. The first concerns problems of diagnosis and classification which are controversial and in a state of flux. It is impossible to do justice to any aspect of the topic without first considering these issues; they are the focus of the initial section of this chapter. The second difficulty arises from lack of research on marital and family treatment of personality disorder. This probably reflects the practical difficulties of engaging the protagonists. Fortunately, there is solid work on the family background of patients and this will be given prominence in what follows. Although the literature suggests a familial contribution, research elucidating the genetic dimension is lacking.

DIAGNOSIS AND CLASSIFICATION

Because the DSM-III-R, which we use here, is similar to the ICD-10, the latter does not require separate consideration. Tyrer's (1990) account of diagnostic snags is elegant and concise. He points out that the various interview schedules take one to three hours to complete, but, with the exception of the borderline and paranoid, correlate at close to zero with clinical assessment. Thus, research is likely to proceed slowly.

Tyrer suggests that the robustness of paranoid personality as a category derives from its clear clinical features: '. . . excessive sensitivity, the tendency to bear grudges and to identify conspiracies, morbid jealousy, and the reading of sinister meanings into neutral events. The enduring nature of borderline disorder as a category is explained by Dahl (1990) who identifies consistent, core features. The most important are disturbances of identity, affect, social and interpersonal functioning, cognitive and perceptual processes, and behavioural organization and impulse control.

In general, the remaining categories appear to have less validity. The reasons for this vary. In the case of schizoid and schizotypal, the problem is the overlap with schizophrenia, the early symptoms of which coincide with the features of schizotypal forms. Similar problems exist with avoidant and dependent, which share many diagnostic features with the anxiety disorders. Although it has been suggested that a comparable overlap occurs between obsessive-compulsive personality and obsessive-compulsive disorder, this is probably not the case (Joffe *et al*. 1988). Thus, it may emerge as one of the more robust categories.

Classification problems with narcissistic and passive-aggressive categories relate to their derivation from psychoanalytic theory. Especially in the case of the former, its theoretical origins are difficult to operationalize and test empirically. Clinically it is quite rare and, together with passive-aggressive disorder, has low diagnostic reliability (Hyler *et al*. 1989).

Antisocial disorder enjoys excellent diagnostic reliability because its features can be verified through external sources such as school and employment records and frequency of law-breaking and debts. However, its validity is undermined by its overlap with features of borderline personality characterized by irritability and irresponsible behaviour. A criticism of the histrionic category is that of sexism. Although its features are probably shared between the sexes, it has traditionally been a category applied to women. A similar gender bias toward men has occurred with antisocial personality.

All these problems, reflected in the relevant literature, make it difficult to deal systematically with the family dimension. The dilemma resolves itself pragmatically since almost all of the work on personality disorder focuses on the borderline personality; this creates a convenient starting point.

BORDERLINE PERSONALITY DISORDER

Much has been written about its developmental origins, most couched in psychodynamic terms. We begin by outlining theories about the family contribution.

Theories of developmental origins

Masterson (1978) presents a lucid view of the developmental origins of the borderline personality. In outlining his theories, which rely on the work of his psychoanalytic predecessors, we include some fairly basic notions. We make no apology for this; without an understanding of the underlying psychological mechanisms, informed family therapy is rendered impossible.

In the first months of life, the infant cannot distinguish between itself and the outside world. When in a state of fear or hunger, unpleasant feelings are projected onto the outside world (primarily represented by the mother) and become part of it, forming the 'bad object'. Pleasant feelings such as those experienced during feeding are projected outward in the same way to form the 'good object'. As the infant becomes aware of the separate existence of the outside world, the bad object becomes increasingly frightening and persecutory. In normal psychological development, a struggle occurs during the latter part of the first year to keep good and bad objects separate. The separation, aimed at protecting the good object, is the origin of the clinical concept of splitting. Positive experiences reduce the persecutory qualities of the bad object and allow the good object to become more robust. This lessens the need for splitting, which is nonetheless reactivated toward the end of the first year when the infant begins to recognize its mother (or other nurturing figure) as a whole object. At this point the infant realizes that it loves and hates the same object. This leads to a fear that aggressive impulses will destroy the good object or that when the good object is absent, it has actually been destroyed. If these fears are unmanageably strong, the development of the infant may become arrested at that stage typified by splitting.

Continuing developmental problems in the second year, which includes the separation-individuation phase, result in the fixation of splitting as fundamental. Because the infant's autonomy is perceived by the mother as a threat, it is discouraged. Instead, regressive, clinging behaviour is rewarded. Splitting becomes elaborated and consolidated around the opposing poles of regressive withdrawal versus the drive toward individuation. These themes are constantly recapitulated in borderline adults, in whom both self and object representations are split. Links between part-self and part-object representations are formed by affective bonds which are so powerful that they preserve the magical thinking and other psychotic aspects of the borderline personality.

Empirical studies of family of origin

The first report of adequate research was by Gunderson *et al.* (1980). These authors compared the parents of 12 carefully diagnosed borderline patients with 12 paranoid schizophrenic and 12 neurotic patients; findings were strikingly consistent. Both mothers and fathers of the borderlines were more psychiatrically disturbed and less functional than parents in the other groups. The borderline families were '. . . characterised by the rigid tightness of the marital bond to the exclusion of the attention, support or protection of the children'. This meant that a poor or non-existent relationship with one parent was not offset by a positive one with the other. Thus, the parental unit failed to provide nurturing or caring. Features most typical of the mothers' adjustment were psychotic in character and included ideas of reference and impaired reality testing, especially denial.

Gunderson *et al.'s* work supports the notion that inadequate nurturing in the first two years contributes to the development of borderline personality disorder. Masterson's ideas about failure of separation-individuation imply maternal overprotection, a feature notably absent from Gunderson *et al.*'s findings. However, Parker (1983) has pointed out that overprotectiveness alone is rare. More common is affectionless control, associated with rejection and neglect. In Gunderson *et al.*'s sample, parental behaviour coincides with Parker's affectionless control, which he demonstrates as pathogenic.

A raised level of psychiatric disorder in parents of borderline patients was a crucial finding that warranted more investigation. This has been undertaken substantially by Loranger *et al.* (1982) who determined types of disorder occurring in first-degree relatives of 83 women with a DSM-III borderline diagnosis. They compared them with the disorders occurring in relatives of 100 female schizophrenic patients and relatives of 100 women with bipolar disorder. Diagnoses were made by independent judges. The borderlines' relatives were 10 times more likely to have been treated for a borderline or closely related diagnosis than relatives of the other two groups. Moreover, borderline relatives were more likely than the schizophrenic patients' relatives to have been treated for unipolar depression. However, the borderline's relatives were not at greater risk for either mania or schizophrenia.

Subsequent work has yielded similar findings. For example, Pope *et al.* (1983) examined first-degree relatives of female borderline in-patients and found a significant increase in prevalence of 'flamboyant' personality disorder compared with the relatives of schizophrenic and bipolar control groups. The borderlines' relatives also had a greater prevalence of major affective disorder than the schizophrenics' relatives. Baron *et al.* (1985) studied relatives of non-patient volunteers who met DSM-III borderline criteria; the incidence of the condition was higher in their relatives than in controls' relatives. Finally, Zanarini *et al.*'s (1988) results also firmly support those of Loranger *et al.* (1982).

Links *et al.* (1988) have determined prevalence of disorder in the first-degree relatives of 69 borderline patients. Since information was sought from direct interviews with relatives, the research is unusually robust. The lifetime prevalence of borderline was 11 per cent. Rates for other disorders were: schizophrenia or schizoaffective disorder, 1.3 per cent; major affective disorder, 16 per cent; substance abuse disorder, 9 per cent; alcoholism, 15 per cent; and antisocial personality disorder, 7 per cent. These rates are similar to those from other work.

Silverman *et al.* (1991) has extended the study of relatives to include affective instability and impulsive personality traits as well as psychiatric diagnosis. The risk for all these were greater in the relatives of borderlines than in those of the control groups.

Clinical studies of family of origin

Although clinical studies lack the rigour of empirical research, they do yield valuable insights. Empirical findings about families of borderline patients have generated a remarkable consensus, but the research has revealed little about the mechanisms through which familial transmission may occur. Important insights into this arise from the work described below.

Mandelbaum (1980) suggests that parents of children (who as adults develop borderline features) use them as targets for projecting their own marital and personal problems. Both parents are enmeshed with their own families of origin, although this is obscured by a distant relationship with them, which is not to be mistaken for differentiation and autonomy. Each generation tends to have a high prevalence of early childhood trauma, originating in parental loss through death, separation, or divorce, usually associated with alcoholism and substance abuse. There is a failure to integrate good and bad objects, so that family members tend to have continuing problem in relationships. Unresolved separation-individuation issues diminish reality testing, and hinder development of basic trust. This manifests as a family pattern of rigid rules and fear of change, which contribute in turn to difficulties in problem-solving.

Feldman and Guttman (1984) base their work on extensive experience of both family and marital therapy with borderline patients. They highlight an interactional pattern which they regard as common in parents of children who developed borderline disorder in adolescence or early adulthood. As part of this pattern, one parent is severely personality disordered, while the other fails to protect the child against the adverse effects of this. Two types of psychopathology follow. In one, the more destructive parent is literal-minded and lacks the ability to empathize and respond to the child's emotional needs. The healthier parent fails to interpret the child's reactions. In the second, a parent has a borderline personality and the child becomes a target of his projections and distortions of reality. The other parent fails to protect the child from these influences and may even form a coalition with the borderline spouse.

The childhood experiences of borderline patients are a potentially rich source of insight into aetiology. However, retrospective biases make data unreliable. Ogata *et al.* (1990) review the literature succinctly and report their findings on 24 borderline in-patients. Typically, the patients recall their parents as disabled by psychiatric disorder, their problems recognized by the children from a young age. Mothers were disabled by depression, fathers by alcoholism and substance abuse. Although parents were recalled as erratic and unstable in their personal functioning and relationships, prevalence of parental loss through death, separation, or divorce was no higher than in a control group. A surprising finding was the high rate of physical and sexual abuse, estimated at 70 per cent and significantly higher than in the controls. Over 60 per cent of patients had witnessed domestic violence.

In summary, Ogata *et al.'s* work suggests that a combination of emotional neglect and trauma, especially physical and sexual abuse, typify the border-line patients' childhood experience. Its negative effects become magnified during adolescence, with difficulties in establishing peer relationships and problems with authority. Poor ego identity contributes to feelings of isolation and separateness from others. Drugs and promiscuity may become the only avenues for emotionally meaningful interpersonal contact.

Family intervention

Much stress has been placed on problems encountered in the individual therapy of the borderline patient. Equally prominent is the difficulty of in-patient treatment (Hartocollis 1980). While it is agreed that effective treatment may require a combination of individual, family, and hospital treatment, clear descriptions of a multimodal approach are scanty. Thus, borderline patients still tend to receive psychotherapy that is exclusively individual with admission during exacerbations of their condition.

Rosenbluth (1987) emphasizes problems of in-patient treatment. We highlight his cogent contribution to demonstrate the need for greater involvement of families. Rosenbluth levels much of his criticism at long-term hospital treatment, suggesting that it promotes regression without assisting the patient to develop shifts in attitude and behaviour that permit greater independence. Thus, after discharge, she frequently seeks readmission during crises, often manifesting self-destructive behaviour if admission is not immediately arranged. This poor outcome is reflected in the few follow-up studies that have been done. Short-term admissions are more likely to be helpful, especially when patient and staff agree on clearly defined goals and readmission criteria. Relapse is also less likely if family members are involved in aftercare plans, which should be comprehensive. As part of this, family members are invited to review their own ways of coping with the patient. However, Rosenbluth's recommendations fall short of systematic marital or family therapy.

It has been claimed that many of the drawbacks of in-patient treatment are avoidable in therapeutic communities. Unfortunately, little evidence is available to support the claim. For this and other reasons, therapeutic communities are rare. Mehlum *et al.* (1991) found that applying thera-peutic community principles in a day hospital helped borderline but not schizotypal patients. This approach may constitute a genuine advance, especially if it incorporates appropriate family involvement.

Little work has been done on the impact of the borderline patient on the family. Schulz *et al.* (1985) attempted to remedy this gap in a study of 31 borderline or schizotypal patients. They do not distinguish between the two conditions in their report. In quantifying the degree of burden on the family, 15 per cent of relatives rated it extreme, 31 per cent high, and 34 per cent moderate. On average, the burden was rated higher than that thought to be involved in managing a person with a physical illness such as cancer, hypertension, or diabetes. However, ratings of burden were considerably lower than those from a previous study of schizophrenic families (Schulz *et al.* 1982).

With regard to specific problems, relatives felt that 'anger as the predominant mood' was most troublesome. They were also concerned about antisocial acts such as drunkenness, substance abuse, promiscuity, and absenteeism. Many emphasized chronic economic difficulties associated with the patient, related to debt and unemployment. Chronic dependency was also stressed. Schulz *et al.* asked similar questions of the patients, but found little agreement about burden between them and relatives. The authors concluded that family education is a useful adjunct to individual therapy and suggested that 'The whole family unit could benefit from identification of and communication about behaviours that cause embarrassment and burden'. However, as in the case of Rosenbluth (1987), they fall short of recommending formal family therapy.

We have noted the scant attention paid to family therapy for the borderline patient. Bloch *et al.* (1991) reviewed 50 families whom they treated in a unit for the family therapy of adult patients. They used a systems approach, including circular questioning to generate hypotheses about family function. These hypotheses were shared with the family and became the main thematic concerns in the therapy sessions. While suggesting that the approach could be helpful in treating personality disorder, this remains to be demonstrated through further research.

Hartmann and Boerger (1989) outline problems of working with families of borderline patients, and discuss ways of overcoming these. Such families may feel blamed, and staff may experience members as demanding and frustrating. Where the family does feel blamed, members are inclined to 'sabotage' treatment. A particular snag is identifying family members suitable for family work. This reflects the propensity of borderline patients to alienate people, including relatives. It is helpful to adopt a broad definition of family in this regard, one embracing extended family members and friends.

Hartmann and Boerger stress the need to keep interactions with the family straightforward and basic. It is especially important not to recruit members as therapists. Family intervention includes setting goals, limited initially and increasing in scope only as treatment progresses and a positive alliance evolves. In general, strategies helpful with the borderline patient are also useful with the family. Although such interventions necessarily confront psychopathology in members, its treatment is addressed elsewhere. Confrontation requires tact to avoid alienating the family from both staff and patient. Limit-setting is important to pre-empt the family making inappropriate demands on the therapists. Helping the family to set limits on the patient's behaviour is equally salient. Most families need help in managing suicidal or homicidal threats; it is important that they take these seriously while at the same time learning to manage them more effectively. Frequently, the need arises for the family to disengage appropriately. Support here is invaluable, especially in advancing the notion that disengagement does not mean abandonment.

Work dealing with formal family therapy for borderline patients is reported rarely. Lansky's (1989) important paper describes an in-patient programme lasting one to three months. It includes an educational component, a psychotherapy group, family intervention, and milieu therapy. Its aims are to develop and sustain an integrated approach to the hurdles of working with borderline patients – splitting, rage attacks, and provocation of either rejection or overprotectiveness in close relatives and hospital staff. The family aspect is systems-based but not restricted to formal therapy sessions alone. Understanding the family dynamics allows staff to interpret the patient's difficulties in a family context. This enhances the potency of their interventions. Emphasis on sustained family involvement also allows a comparison of the patient's support systems: in childhood, on the ward, and in the current family. The ward and family both function as 'containers' and share a responsibility for the patient's welfare. Comparison of containment systems helps to integrate various modes of treatment. Family therapy is often intensive but the risk of negative repercussions is reduced by the safe hospital environment. Skilled staff deal with acting out as well as with issues that cannot be addressed during specific family sessions.

Marital therapy

Although there are no reports on marital therapy specifically for the borderline patient, Weddige (1986) identifies the major issues by focusing on individual therapy for those who continue to live with their borderline spouses. Because the approach aims at helping the spouse cope better with the marital relationship, it is, in effect, a form of marital therapy. Weddige stresses the following:

1. Self-doubt – spouses blame themselves for contributing to the patient's problems, and feel guilty, inadequate, and depressed. They may re-examine the past scrupulously and repeatedly in an attempt to make

sense of current difficulties, believing that the situation might be better if they had acted differently.

2. Personalizing and reaction-formation – when incessantly confronted with negative projections, the spouse comes to believe what is said, and this promotes self-doubt. Anger about these projections or about particularly stressful marital episodes is often dealt with using reaction-formation, whereby the spouse becomes exceptionally devoted, excelling with acts of kindness.

3. Denial, projection, and introjection – spouses deny the full extent and nature of the problem. They may project negative feelings onto friends or coworkers, creating interpersonal difficulties that increase their sense of isolation, which often becomes profound. Frequently, spouses rely on introjection, suppression, and repression to preserve equilibrium, especially in the face of repeated angry outbursts. This pattern of defence often leads to chronic depression.

Effective therapy for these spouses calls for an understanding of the above difficulties. It may be necessary to interview the borderline partner before this understanding can be achieved. Educating the spouse about the nature of the borderline is crucial since it helps to reduce self-blame and to facilitate the accomplishment of more adaptive coping strategies. For example, the spouse learns of the patient's need for space and ego-boundaries, as a result of which issues around intimacy may be better managed. During this process, the spouse needs support and encouragement together with informed guidance about how to deal with specific issues. It is particularly important to teach him how to become the children's advocate without alienating the borderline spouse.

Although Weddige does not advocate it himself, there appear to be no major impediments to applying a conjoint marital approach. Obviously, this requires skill and tact, and will often need to be implemented over an extended period. Inevitably, some sessions will be stormy and stressful. Thus, a commitment to therapy is needed from all those persons involved. Given this, a couples approach might yield benefits not only in the spouse but also in the borderline partner.

Support for a conjoint approach comes from a book by Lachkar (1992). This work is concerned with therapy where both partners are personality disordered, one narcissistic, and the other borderline. The partner with the former is considered to be the healthier one. Lachkar provides no data on prevalence of these marriages, which is probably not high. None the less, her book is so rich in clinical insights that we think it useful to summarize her guidelines:

1. Because conjoint therapy may diminish a sense of individuality, it is vital to provide couples with clearly defined personal boundaries. The more disturbed the couple, the more is such structure required. The need to illustrate the nature and extent of regressive behaviour is also important.

2. The narcissist's experience of anxiety differs from that of the borderline; the former is most threatened by loss of a sense of specialness, the latter by a fear of abandonment or fragmentation.

3. The therapeutic alliance is initially established with the narcissistic partner. This counters his tendency toward withdrawal. During this process it is crucial to give sufficient emotional support to the borderline partner.

4. Often, the conjoint approach should be viewed as preparation for individual therapy which may be possible only after issues of dependency and individuation have been addressed conjointly.

SCHIZOTYPAL PERSONALITY DISORDER

Although much less work has been done on the schizotypal than the borderline patient, information has accumulated, including some on family factors. Since the family oriented reports are aimed primarily at distinguishing schizotypal from borderline, this aspect is a logical starting point for our discussion.

Baron *et al.* (1985) sought to test the hypothesis that a familial link exists between schizotypal and schizophrenia on the one hand, and between borderline and depressive illness on the other. Their study was based on 20 patients with a schizotypal diagnosis alone, 16 with both schizotypal and borderline, and 17 with only a borderline diagnosis. A control group of normal subjects was also included. Nearly 600 first-degree relatives was studied; given that 70 per cent were interviewed directly, this study is remarkable for its methodological rigour. The results are clear-cut. Relatives of the schizotypal group had a greater risk for that condition than the other relative groups, but not for schizophrenia. The relatives of the borderlines were at greater risk for that condition, and for major depression.

Baron *et al.* discuss their findings in the context of a literature review and conclude that both schizotypal and borderline disorder are familial. However, they appear to be distinct with little overlap (although they may co-exist). The lack of a familial relationship between schizotypal and schizophrenia, they suggested, could be explained on statistical grounds. The evidence suggests that such a relationship does exist but because of the rarity of combined schizotypal/schizophrenia, large-scale studies are needed.

Schulz and his colleagues (1989) examined similar issues using a different approach. They studied the first-degree relatives of 'pure' borderlines and in-patients with both borderline and schizotypal features. These were compared with a third group comprising relatives of in-patients with schizophrenia. Comparing the relatives of the 'pure' borderline group with those of the mixed group, they detected schizophrenia only among the relatives of the latter. Combining all personality disorder patients, their relatives had higher rates of depression, alcoholism, and antisocial personality than relatives of the schizophrenic patients. The results support the relationship of borderline and the affective spectrum. In addition, severe forms of concurrent borderline and schizotypal are related to the schizophrenic spectrum.

The difference between Baron *et al.* (1985) and Schulz *et al.* (1989) concerns the question of a mixed borderline/schizotypal disorder. Baron *et al.* favour the idea of separate entities that may co-exist. Schulz *et al.* argue for a single disorder that combines features of both borderline and schizotypal: it is this disorder, they suggest, that has the closest connection with schizophrenia.

Although these matters are diagnostically cogent their clinical relevance remains modest. This is underlined by Schulz *et al.*'s 1985 report previously discussed, in which they consider the family impact of both borderline and schizotypal disorder in similar terms. Moreover, their recommendations for family intervention assume that the two conditions are clinically similar. As our knowledge evolves, it may become possible to formulate interventions that take differences between disorders into account. Current knowledge does not permit such a distinction. This is reflected in the dearth of useful reports about family or marital interventions in personality disorders other than borderline or schizotypal.

CONCLUSION

Basic ideas about personality disorder are being challenged by research. From a clinical perspective, the idea of the borderline as a heterogeneous category is important. In the context of family intervention, it suggests that techniques appropriate for borderline patients apply to personality disorders more generally. Whether or not this is true can be resolved only by further research. Given the nosological complexity it will take several years before such studies are feasible. Meanwhile, clinicians must rely for guidance mainly on clinical knowledge. Although the few reports of family intervention in personality disorders are encouraging, more are needed urgently (Clarkin *et al*. 1991).

REFERENCES

Baron, N., Gruen, R., Asnis, L. *et al.* (1985). Familial transmission of schizotypal and borderline personality disorders. *American Journal of Psychiatry*, **142**, 927–34.

Bloch, S., Sharpe, M., and Allman, P. (1991). Systemic family therapy in adult psychiatry. *British Journal of Psychiatry*, **149**, 357–64.

Clarkin, J.F., Marziali, E. and Munroe-Blum, H. (1991). Group and family treatments for borderline personality disorder. *Hospital and Community Psychiatry*, **42**, 1038–43.

Dahl, A.A. (1990). Empirical evidence for a core borderline syndrome. *Journal of Personality Disorders*, **4**, 192–202.

Feldman, R.B. and Guttman, H.A. (1984). Families of borderline patients: Literal-minded parents, borderline parents, and parental protectiveness. *American Journal of Psychiatry*, **141**, 1392–6.

Gunderson, J.G., Kerr, J., and Englund, D.W. (1980). The families of borderlines. *Archives of General Psychiatry*, **37**, 27–33.

Hartmann, D. and Boerger, M.J. (1989). The families of borderline clients: Opening the door to therapeutic interaction. *Perspectives in Psychiatric Care*, **25**, 15–17.

Hartocollis, P. (1980). Long-term hospital treatment for adult patients with borderline and narcissistic disorders. *Bulletin of the Menninger Clinic*, **44**, 212–26.

Hyler, S.E., Reider, R.O., Williams, J.B. *et al.* (1989). A comparison of clinical and self-report diagnoses of DSM-III personality disorders in 552 patients. *Comprehensive Psychiatry*, **30**, 170–8.

Joffe, R.T., Swinson, R.P., and Regan, J.J. (1988). Personality features of obsessive-compulsive disorder. *American Journal of Psychiatry*, **145**, 1127–9.

Lachkar, J. (1992). *The narcissistic/borderline couple. A psychoanalytic perspective on marital treatment*. Brunner/ Mazel, New York.

Lansky, M.R. (1989). The subacute hospital treatment of the borderline patient: II. Management of suicidal crisis by family intervention. *Hillside Journal of Clinical Psychiatry*, **11**, 81–97.

Links, P.S., Steiner, M., and Huxley, G. (1988). The occurrence of borderline personality disorder in the families of borderline patients. *Journal of Personality Disorders,* **2**, 14–20.

Loranger, A.W., Oldham, J.M., and Tulis, E.H. (1982). Familial transmission of DSM-III borderline personality disorder. *Archives of General Psychiatry*, **39**, 795–9.

Mandelbaum, A. (1980). Family characteristics of patients with borderline and narcissistic disorders. *Bulletin of the Menninger Clinic*, **44**, 201–11.

Masterson, J. (1978). *New perspectives on the psychotherapy of the borderline adult*. Brunner/Mazel, New York.

Mehlum, L., Friis, S., Irion, T. *et al.* (1991). Personality disorders 2–5 years after treatment: A prospective follow-up study. *Acta Psychiatrica Scandinavica*, **84**, 72–7.

Ogata, S.N., Silk, K.R., and Goodrich, S. (1990). The childhood experience of the borderline patients. In *Family environment and borderline personality disorder* (ed. P.S. Links). American Psychiatric Press, Washington.

Parker, G. (1983). *Parental overprotection*. Grune and Stratton, New York.

Pope, H.G., Jonas, J.M., Hudson, J.I. *et al.* (1983). The validity of DSM-III borderline personality disorder. *Archives of General Psychiatry*, **40**, 23–30.

Rosenbluth, M. (1987). The impatient treatment of the borderline personality disorder: A critical review and discussion of aftercare implications. *Canadian Journal of Psychiatry*, **32**, 228–37.

Schulz, P.M., Schulz, S.C., Dibble, E. *et al.* (1982). Patient and family attitudes about schizophrenia: implications for genetic counselling. *Schizophrenia Bulletin*, **8**, 504–13.

Schulz, P.M., Schulz, S.C., Hammer, R. *et al.* (1985). The impact of borderline and schizotypal personality disorders on patients and their families. *Hospital and Community Psychiatry*, **36**, 879–82.

Schulz, P.M., Soloff, P.H., Kelly, T. *et al.* (1989). A family history of borderline subtypes. *Journal of Personality Disorders*, **3**, 217–29.

Silverman, J.M., Pinkham, L., Horvath, T.B. *et al.* (1991). Affective and impulsive personality disorder traits in the relatives of patients with borderline personality disorder. *American Journal of Psychiatry*, **148**, 1378–85.

Tyrer, P. (1990). Diagnosing personality disorders. *Current Opinion in Psychiatry*, **3**, 182–7.

Weddige, R.L. (1986). The hidden psycotherapeutic dilemma: Spouse of the borderline. *American Journal of Psychotherapy*, **40**, 52–61.

Zanarini, M.C., Gunderson, J.G., Marino, M.F. *et al.* (1988). DSM-III disorders in the families of borderline outpatients. *Journal of Personality Disorders*, **2**, 292–302.

12

Dementia and the family
Henry Brodaty

Age takes hold of us by surprise
(Goethe)

Dementia has attracted significant attention in recent years for good reasons. In the twentieth century, for the first time ever, a person can expect to live past the age of retirement, 65 years. Increasing longevity and falling birth rates have resulted in demographic patterns never before experienced. Not only are there more old people, those surviving into the senium are dying more often of degenerative diseases such as dementia, while death from infection, heart attack, and stroke are waning.

Concomitantly, researchers have discovered that old age provides fertile ground to till. The seminal paper by Roth (1955) led clinicians to abandon the senile label previously applied to all older people with mental illness. Paranoid states, affective disorders, and acute and chronic organic brain syndromes were delineated by their different longitudinal courses and outcomes, and, with the arrival of psychotropic medications, by different treatments. Under closer scrutiny, the dementias came to be understood more clearly. Blessed *et al.* (1968) found that the degree of senile dementia correlated with neuropathological changes characteristic of Alzheimer's disease, senile plaques, and neurofibrillary tangles, and not with the degree of atherosclerosis as had previously been thought. Suddenly a disease that had been considered uncommon and confined to the presenium (that is, before the age of 65 years), became an epidemic. Nursing homes were no longer filled with 'senile dements' but with sufferers of an interesting condition, Alzheimer's disease.

This new found medicalized respectability for the dementias was reinforced by the rise of self-help groups – Alzheimer's Disease and Related Disorders Societies and Associations – in the late 1970s and early 1980s in England, the United States, and Australia and thereafter throughout the world. These consumer-based organizations generated enormous publicity and by the late 1980s Alzheimer's disease had become a household word. The increasing fear of dementia has become manifest by a spate of jokes, films, and books on the topic. Alzheimer-phobia has come to rival cancer-phobia.

The epidemic of dementia has been paralleled by changes in family structure in Western countries namely urbanization and the decline of the extended family. As a consequence there has been a proliferation of institutional care and residential facilities and more latterly professional community services. The balance between these and family care has varied according to culture, country, and economics. It is against this background that the interaction of dementia and the family is examined.

THE NATURE OF DEMENTIA

Dementia, the loss of self (Cohen and Eisdorfer 1988), is a syndrome of organic aetiology characterized by impairment of memory and other intellectual functions, associated with change in personality. The first signs are almost imperceptible and its progression is variable but most commonly insidious. The affected person may wonder whether his increasing forgetfulness is merely part of normal ageing or whether it is the sinister harbinger of a more malign disorder. The family may be at a loss to explain the subtle but strange changes in personality and behaviour that often herald the onset of dementia. As the condition progresses the balance of relationships changes, forever.

Dementia of the Alzheimer's type is the most common cause of dementia, representing approximately 50 per cent of the dementias. A further 15–20 per cent are due to vascular conditions – multi-infarct dementia, Binswanger's disease, and strokes and about 15–20 per cent are caused by the combination of vascular and Alzheimer-type degeneration. Alcohol and head injury are the next most common causes followed by a host of other conditions, of which only few are potentially reversible, for example, thyroid disease, vitamin B12 deficiency, and normal pressure hydrocephalus. The clinical features and their effects on families of these various types of dementia may differ, mostly subtly, sometimes significantly. For example Pick's disease commonly presents with behavioural problems before memory difficulties emerge. Seemingly inexplicable alterations in personality and patterns of interaction may cause serious family disruption and extensive psychiatric consultations before the organic aetiology is realized. For this chapter, Alzheimer's disease provides a convenient prototype for considering the effects of dementia on the family.

THE PREVALENCE OF DEMENTIA

As age is the major risk factor for dementia, the worldwide trend for the 'greying' of the population is of obvious importance. In Australia, New Zealand, and the United States there will be a doubling of the proportion

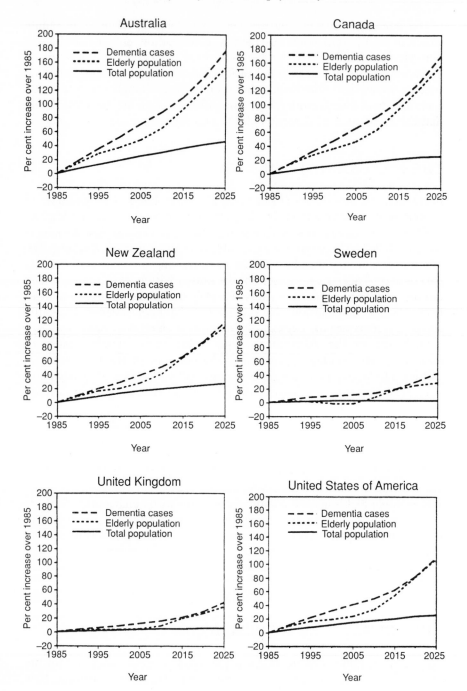

of people aged 65 years and older over the next 40 years, from the current 11 per cent of the total population to about 20 per cent (see Fig. 12.1) (Jorm and Henderson 1990). This trend is particularly apparent in the very old, those over 80 years, who are the fastest growing sector of the population, and in developing countries which have low proportions of older people currently and are set to have dramatic percentage increases in coming decades.

About 5 per cent of people over the age of 65 have moderate to severe dementia, 20 per cent of those over 80 and 30 per cent of those over 90. Younger people are also affected by dementia: for every nine people over the age of 65 so affected there is one person younger (US Congress, Office of Technology Assessment 1987). They and their families face extra problems.

Thus in 1992 in the USA there were approximately four million people with moderate to severe dementia, in the UK about one million and in Australia about 120 000 (Jorm and Henderson 1990). (Prevalence figures for mild to moderate dementia are much less certain, as they pose less burden on families and support services and are not usually discussed. Estimates are that there are at least as many people with mild dementia as there are with moderate to severe dementia). In Australia about half those with dementia reside in institutional settings – over 40 000 in nursing homes, 10 000 in hostels and 3 000 or more in hospitals and other facilities. The remainder are in the community.

Another demographic aspect of the impact of dementia on the family is the increasing time span between generations. With women often delaying childbirth to their thirties it is quite common for 40–50-year-old women to be caught between the demands of their own growing families and a dementing parent.

THE 'CAREER' OF THE DEMENTIA

Chronic illnesses are rarely static. A person is dementing rather than demented. The clinical course, almost always relentlessly progressive and downhill, though interpersed occasionally by plateaus, can be conveniently divided into three stages.

Early, the family struggles to recognize the problems, gain an assessment

---

Fig. 12.1 Projected increases in dementia cases, elderly population, for six countries, 1985–2025. Reproduced with permission of Dr. Jorm and taken from Department of Community Services and Health (1990). *The problem of dementia in Australia.* Second edn. by A.F. Jorm and A.S. Henderson. Australian Government Publishing Service, Canberra.

and a diagnosis, and adjust to the news. Planning is important as practical, financial, and legal consequences follow. Contingency procedures should be arranged in case of mishap or acute illness in the carer or another emergency occurs. Progression of dementia is associated with the loss of the abilities to understand concepts, make calculations, read, write, and eventually, sign one's name. The point will come on the downward decline beyond which it will be impossible for the dementing person to operate a bank account, realize assets, or write a will. Preparations before this stage is reached can prevent subsequent problems.

Testamentary capacity is defined similarly in the UK, Australia, and the USA. In English law, sound testamentary capacity means that three things must exist at one and the same time:

(1) 'the testator must understand that he is giving his property to one or more objects of his regard;

(2) he must understand and *recollect* the extent of his property;

(3) he must also understand the nature and extent of the claims upon him both of those whom he is including in his will and those whom he is excluding from his will. The testator must realise that he is signing a will and his mind and will must accompany the physical act of execution. It is not sufficient that the testator be of memory when he makes his will to answer familiar and unusual questions, but he ought to have a disposing memory, so that he is able to make a disposition of his lands with understanding and reason' (Sherrin *et al.* 1987; p. 26, our emphasis).

Further the testator must not be under the influence of delusions which could bring about a disposal of his property which otherwise he would not have made nor subject to undue influence. The latter is a more complex and less precisely defined concept usually containing an element of coercion, compulsion, or restraint (Spar and Garb 1992). Where dementia has been diagnosed and a will is to be written, it is prudent for testamentary capacity to be ascertained formally and documented medically and legally. Guidelines for psychiatric assessment are provided by Spar and Garb (1992). Couples with 'mirror wills', where each bequeathes the entire estate to the other, need to allow for the situation where the well partner predeceases the dementing one.

Power of Attorney is a formal document by which one person (called the principal) empowers another (called the attorney) to act on the principal's behalf, either in relation to the principal's affairs, generally, or in respect of specific matters only (Porter and Robinson 1987). A person of unsound mind has limited power to authorize an attorney. This power does not authorize the attorney appointed to do any act, the nature of which was beyond the understanding of the principal when the power was given (Porter and Robinson 1987, p. 90). There is a fundamental common law principal that Power of Attorney, even if validly created, will be revoked and made invalid by the later mental incapacity of the principal. In several countries and states of Australia it is possible for the principal to specify that the Power of Attorney should endure even if the principal's mind becomes unsound subsequently.

In the middle stage of the dementia the problems of declining cognitive powers are accompanied by increasing demands on the carer, by the affected person's increasing dependency, and by more difficult problem behaviours such as aggression and wandering. Carers complain bitterly about the loss of a companion and the absence of someone with whom to communicate. Delusions, hallucinations, misidentification, and depression may add further complications.

In the last stage the dementing person may require assistance with basic activities of daily living such as dressing, bathing, and toileting and may become doubly incontinent. Families are often consumed with guilt when they, unable to provide the high level of nursing care required, are forced to arrange admission of their relative to a nursing home.

THE FAMILY AS CARERS

Dementia, although an 'unremitting burden' on the family (Anderson 1987) is best cared for by a spouse or relative, for without her or him the dementing person is more likely to be placed, and more quickly placed, in residential care. Even for the 10 per cent of people with dementia who live alone families often remain involved, albeit at a distance. Furthermore, institutional placement, when it occurs, may change the nature of family involvement but certainly does not end it.

Spouses, particularly wives twice as often as men, are the most frequent carer accounting for about 60 per cent of domestic carers (Wells *et al.* 1990). About one in three carers is a child, almost always the daughter (or daughter-in-law) of dementing parents. Other relatives and friends make up the remainder. About 75 per cent of carers are women (Brodaty and Hadzi-Pavlovic 1990) whom society traditionally views as responsible for sick family members. Thus a daughter rather than a son is more likely to assume care of a dementing parent, or if no daughter is available it may be the daughter-in- law. Brody (1981) has drawn attention to these 'women in the middle' caught between competing demands of spouse, their own children, dementing parents or parents-in- law, and often their own career. Sons are less frequently the supporters of dementing parents and their role has received little attention.

PATTERNS OF CARE

Care-managers and care-providers

Families care in different ways. Some are care-providers – usually spouses, especially wives; others are care-managers – especially men, and the children of the dementing person. Care-providers feed, wash, and spend the days with their dementing partners. Care-managers on the other hand, busy with family and work, organize: Meals-on-Wheels, a domiciliary nurse, a personal care assistant to wash the impaired person, or a relative to accompany a dementing parent. Care-managers as a rule are more likely to set limits, delegate, and take an objective, instrumental approach to tasks (Colerick and George 1986).

The consequences of these patterns are that care-managers are less personally involved and have less psychological distress. Care-providers suffer more distress but wait much longer before relinquishing care. Thus the children are more likely to be considering residential placement of their dementing parent at a stage when a spouse might just be seeking support services.

Reasons for caring

What sustains carers for the long haul, the 36-hour day (Mace and Rabins 1981), the 8-day week, year-in year-out? Eisdorfer (1991) postulated four principal motivations: love, equity (called reciprocity by others), morality (that is, societal and cultural expectations), and rarely, greed. Love and affection are often supported by the belief that 'no one can care as well' as the carer. There may be spiritual or altruistic elements intermingled as carers, unshackled by ambition and striving, are now able to extract the most out of the limited number of remaining days. These carers accept the ineluctable fate of the dementia and achieve harmony by living for the moment rather than dwelling on future hardship. This contrasts with another pattern frequently seen in some carers who deny the inevitable and are sustained, for a time, by a constant, frantic though futile search for miracle cures.

Equity or reciprocity is often exemplified by such statements as 'she looked after me for so many years . . .' or, 'he'd do the same for me'. Some carers freely repay credits built up over a lifelong relationship. Others do so grudgingly, yoked together by guilt and obligation, but are demonstrably more distressed (Gilleard 1984). It has been suggested that men assume the caring mantle more readily as a change from 40 or 50 years of job-related activities. Women, jaded by a lifetime of providing nurture to children, husband, parents, and grandchildren are more likely to feel cheated out of their retirement. Some carers obtain a perverse pleasure in refusing all offers of assistance.

Their resolve redoubles with each dire warning of their imminent collapse. Such martyrdom may be a defence against the pain of grief and the anxiety of future isolation. Those carers with obsessional traits may derive satisfaction from completing the tasks of caring, albeit mechanically and devoid of tenderness.

Societal expectations may be embodied in family beliefs. There is a clear hierarchy of expectations about which family member will take on the caring role – (in order) partner, unmarried daughter, married daughter, son (including, and sometimes principally his wife), and distant relative (Hall 1990). Expectations and beliefs can exert intense pressures, particularly on women. Brody (1985) found that over half of all working women who had to care for an elderly relative had quit work or altered their pattern of employment. The media may have had a role in reinforcing guilty feelings by propagating societal expectations that families keep a dementing person at home at all costs.

CARER STRESS

Effects on carers

It is clear that caring for a person with dementia is stressful. Indeed caring for any chronic mental condition causes more stress than caring for a physical disorder (Lezak 1978). The determinants of how carers cope and the effects on them are complex: the nature of the illness, the carer's personality and coping skills, the quality of the relationship before the illness, and, to a lesser extent, the impact of the extended family and society in general all play a part.

Carer distress

A plethora of studies document the high rates of carer distress (Gilleard 1984; Anderson 1987; Brodaty and Hadzi-Pavlovic 1990; Gilleard 1992) whether measured by scores on the General Health Questionnaire (GHQ) (Goldberg 1972), Beck Depression Inventory (BDI) (Beck *et al.* 1961), or other scales of psychological morbidity. A useful model for understanding this distress, proposed by Poulshock and Deimling (1984), is that there is a burden which is objective and measurable (for example, incontinence, frequency of aggressive behaviour, number of hours of personal care) which causes stress on the carer. The effects of this stress on carers is called the strain, that is, subjective distress which might be measurable on the GHQ or BDI. Thus some carers may be able to cope fairly well and not exhibit any strain while others, unable to tolerate the same level of stress, become depressed.

There is a dearth of longitudinal data to help understand the effects on the carer. It is clear that the severity of the dementia and the duration of caregiving are not associated with the degree of distress. Surprisingly, the degree of dependency and wandering seems to have little effect on carers. On the other hand apathy and withdrawal as well as aggressive, disturbing, and demanding behaviours are highly correlated with high GHQ scores in carers (Brodaty and Hadzi-Pavlovic 1990; Draper *et al*. 1992; Gilleard 1992). That is, the day-to-day grind of dressing and feeding someone, of helping a spouse with the bath, and of providing for menial wants is a managerial task that is physically demanding but not necessarily an emotional strain. It is the loss of communication, the barrage of constant, repeated, and stereotyped questioning, the need for constant supervision, and the change in personality that accompanies dementia, which are extremely distressing. Not only do spouses often complain that they have lost the person they loved but, instead, they have gained an unpredictable, demanding new person whom they do not like very much.

Ironically, at the very time that carers are most distressed they are often most isolated. Friends and family, who presumably feel awkward, embarrassed, and uncertain, visit less often. Carers, trapped by concern for the safety of a dementing person left alone, journey out of the house less and less frequently. Their distress may make others uncomfortable, further sealing them in their isolation.

Studies examining physical morbidity in carers have been reviewed by Schulz *et al*. (1990). Carers have reported themselves as having poorer health than controls and having more chronic illness. In one study they attributed this as being caused by the support role. Health care utilization studies have been inconsistent, with some investigators reporting more frequent physician visits and more prescription drug use by carers than matched controls while others did not. A possible confound may be that the demands on carers may have limited their opportunities to seek professional health care. Our own survey found that psychological distress, physical ill health, and drug use tended to occur together (Brodaty and Hadzi-Pavlovic 1990). In one study, immune function was found to be impaired in carers of Alzheimer's disease patients as compared with controls and this was independent of nutrition, sleep, or other health-related behaviours (Kiecolt-Glaser *et al*. 1987). While the carers in that study had higher levels of depression, lower life-satisfaction, and poorer mental health, the authors did not report whether there were any correlations between psychological measures and immune function (Schulz *et al*. 1990).

For 'the carer to decide to give care s/he must also decide not to care about some other area of life – be it work, other relationships or simply one's independence' (Aronson 1990). This decision to give up work has obvious financial consequences, as may the necessity to employ a personal care assistant or to arrange for residential care. In one USA study, carers were more likely than their age peers to report adjusted family incomes below the poverty line (Stone *et al.* 1987).

The change in the relationship

The shift in the balance of relationships consequent on care-giving are painful. The dependent older person becomes anxious about being a burden. He may feel increasingly inadequate, unattractive, and unwanted and project these attitudes onto an innocent spouse who perhaps returns home after an hour's shopping only to be accused of infidelity. His resentment at his loss of autonomy and independence may manifest as hostility and irritability. For the carer, whether she be a daughter, a wife, or another relation, the shift may lead to role conflict. Cast in the role of surrogate mother, the carer feels trapped and resentful, but quickly suppresses negative thoughts, guiltily chastising herself.

Marital relationships usually deteriorate. As a rule the dementing person's interest in sexual activity diminishes although occasionally the reverse occurs, almost always in younger (less than 60 years old) dementing men, often associated with frontal lobe pathology and disinhibition. Ironically, the companion with whom the carer would have discussed her feelings is now the cause of them and cannot respond meaningfully.

The distress resulting from this loss of a companion, simultaneously psychologically absent and physically present, can be better understood through the concept of boundary ambiguity (Boss *et al.* 1990). This is a state in which 'family members are uncertain of who is in or out of the family and who is performing what roles and tasks within that system'. The concept was developed from work with military wives whose husbands were missing in action and therefore physically absent but psychologically present. In dementia the reverse occurs. In the face of high ambiguity, change is blocked and grieving cannot be completed. Incomplete, prolonged, even continual mourning for the lost parts of the affected person and the lost relationship, may lead to a chronic depression in the carer.

THE VULNERABLE CARER

An abundance of studies (Zarit and Toseland 1989) have confirmed that carers at risk of being significantly distressed, decompensating, or being unable to provide continuing care can be identified. Carer stress is associated with variables characterizing the dementing disorder, the affected person, the carer and their relationship.

Dementia variables

The evidence that the severity of the dementia and specific problems associated with dementia affect the level of carer distress is complex. There is probably no correlation between carer distress and dementia disability, severity of memory impairment, or total number of behaviour problems (Morris *et al.* 1988; Brodaty and Hadzi-Pavlovic 1990). Nor is it likely that duration of dementia has a relationship with distress (Brodaty and Hadzi-Pavlovic 1990). Two explanations for this latter finding are suggested. The adaptation hypothesis suggests that as carers become more accustomed to the daily grind of care-giving they become less distressed. The second, the sequestration hypothesis, suggests that the more distressed carers have already placed their dependents in institutions and so do not appear in surveys of carers living with dementing persons.

Certain types of dementia-related behaviour are associated with distress (Gilleard *et al.* 1982; Greene *et al.* 1982; Morris *et al.* 1988, Brodaty and Hadzi-Pavlovic 1990). Incontinence, immobility, nocturnal wandering, proneness to falls, inability to engage in meaningful activities, difficulties with communication, and sleep disturbance have been reported as contributing to carer distress (Morris *et al.* 1988). In general, behaviours which emphasize the loss of companionship, such as apathy and non-communicativeness, and those which are disruptive, such as constant demands or disturbances, are more stressful to the carer than the simple dependency accompanying progressive debilitation.

Carer variables

Carers are predisposed to psychological distress by personality, coping, interpersonal, and socio-environmental and cultural variables as well as age and gender. Carers more likely to be distressed are women rather than men, spouses rather than children, care-givers who are physically unwell and socially isolated (Brodaty and Hadzi-Pavlovic 1990). Also stress is greater in carers who are closer, be it a more immediate blood relationship, a more committed role relationship, or physical propinquity, for example living together.

Effective coping strategies can mitigate carer distress (Pruchno and Resch 1989) and probably enhance the quality of life of the dementing person. Lazarus and Folkman (1984) categorized coping into emotion-based or problem-solving strategies. Emotion-based responses include wishfulness ('wished you were a stronger person to deal with it better', 'wish you could change what had happened', 'wish you could change the way you felt'); acceptance ('accept the situation', 'refuse to let it get to you', 'make the best of it',); and, fantasy or intrapsychic (fantasies about how things could turn out, 'told yourself things to help you feel better'). Problem-focused strategies, which include reframing, problem solving, developing more social support or a greater social network are associated with positive effects such as greater satisfaction with life, decreased feelings of burden, or lower depression levels (Pruchno and Resch 1989). In general, more immature coping strategies are associated with increased burden, decreased satisfaction with life and increased depression (for example, Pruchno and Resch 1989).

Such findings lend support to the view that the psychological characteristics of carers manifested in coping styles and expressed emotion, influence their stress over and above the objective burden experienced. They also suggest that cognitive and/or behaviour therapies aimed at developing more effective coping styles may be beneficial to carers (for example, Brodaty and Gresham 1989).

Caution must be taken in interpreting the literature investigating coping styles. The effects of different patterns of coping cannot necessarily be extrapolated from one type of stress (for example, caring for a diabetic child) to another (for example, coping with a dementing spouse). Nor is it possible, without longitudinal studies, to rule out the possibility that depression may lead some carers to adopt less effective types of coping. It may be that depression itself leads to the use of less effective forms of coping and this in turn causes more depression – the interactive model.

Relationship factors

The influence of the pre-existing relationship on carer distress is not clear. Generally, evidence suggests that those in unsatisfactory relationships before the onset of the dementia are more likely to be distressed subsequently. Not surprisingly, there is a strong association between deterioration of the carer's perception of the quality of the marital relationship and distress (Brodaty and Hadzi-Pavlovic 1990). It is difficult to evaluate these findings as carer perceptions of past and present relationships are likely to be heavily confounded by current mood state. Yet it is easy to understand how distressing it must be to families for as communication becomes fragmented, tension increases (Chenoweth and Spencer 1986), companionship and a confidant are lost (Fitting *et al.* 1986; Wright, 1991), economic and household responsibilities shift more and more onto the carer's shoulders, and sexuality wanes (Wright 1991). There are often considerable discrepancies between the views of the marriage of the affected person and the carer; the latter usually being more negative (Wright 1991).

Recent research on the affective quality of the relationship within families measured in terms of Expressed Emotion (EE) reported associations with strain and distress in daughters of dementia sufferers (Bledin *et al.* 1990). Women who were rated high on EE (high levels of critical comments and/or hostility toward patient) were more likely to report more strain and distress, although they were no different from low scoring EE daughters in their use of maladaptive or positive coping strategies and there was no difference in severity of dementia between the two groups. It is unclear however, whether EE is causally related to strain or simply an expression of strain in the carer.

The feelings and thoughts of the dementing person have been largely neglected by researchers. Bewilderment, frustration, anger, irritability, anxiety about the future and about being a burden on the family, fear about being 'put away', and depression are common emotional reactions which in turn have effects on the family. The feelings of each partner inevitably influence the other, sometimes setting up a feedback loop of anger and resentment too.

Support

There is still debate as to which supporters are the most helpful for carers (Fiore *et al.* 1986). There are differences between actual and perceived support, formal (from professionals) and informal (from friends and relatives) support, and between males and females in how they use support. There is also potential for negative effects from so-called supportive figures (Fiore *et al.* 1986; Edward and Cooper 1988). For example, the level of support is sometimes measured by the actual number of people coming into the house, yet some visitors may be stressful rather than helpful.

Overall, wives who do not work receive the least help and although husbands and wives spend the same time caring, husbands receive more help (Enright 1991). Whilst help with instrumental tasks may reduce the physical demands on the carer, it does not necessarily reduce the psychological strain experienced by the carer (Lawton *et al.* 1991). It appears therefore that both psychological and instrumental support are required to effectively reduce the strain of caregiving.

Interpreting correlations between supports and distress is fraught with danger. An association between the number of supports and caregiver distress might merely reflect an increased likelihood of distressed carers seeking help. A negative correlation between supports and distress could suggest that support has an ameliorating effect. A lack of either positive or negative correlation might result where both effects occur in the one population sample, that is, for some carers support is helpful whereas for others support is a reflection of their neediness. Finally, there may be a relationship between the use of informal and formal services so that carers receiving more informal help may need less professional help (Fitting *et al.* 1986) and make fewer demands on community services (Chesterman *et al.* 1987).

In summary there is support for a relationship between informal supports and lower psychological distress. That is, the carers who have contact with more people on a daily basis feel better than those who do not. Obviously no causality can be imputed – it may be that happier people are more inclined to seek contact in the first place or that lack of happiness and social support are independently related to a third factor such as nuisance behaviours accompanying the dementia or to the type of society in which care-giver and sufferer live. There is even less evidence that formal supports are helpful; although an English study provided longitudinal evidence of efficacy, only 31 per cent of clients had dementia (Challis *et al.* 1987).

FAMILIES WITH SPECIAL PROBLEMS AND DOUBLE ISOLATION

Migrant familes

Migrant families struggling with dementia are faced with additional predicaments. As dementia progresses, intellectual functions, including language, regress. Languages acquired more recently are forgotten sooner. The dementing person now increasingly reverts to his mother tongue and unknowingly mixes languages together. He finds conversation much more of an effort. Nurses and other helpers cannot communicate as effectively with the affected person.

Prevailing cultural values may be at odds with imported ones. For example, the migrant family may steadfastly resist community or respite services and view nursing home placement as a shameful failure. Assistance with bathing or toileting by a nurse of the opposite sex is often anathema to the elderly person with dementia, particularly where he is from another culture. Professional helpers, even if accepted by the family, are rarely of the same background or culture and thus find it difficult to understand the nuances of family interaction.

Family carers isolated by language, culture, and illness, are also impeded from reaching out. In many countries such problems have already become apparent: in the UK with its waves of Indian, African, and Carribean migrants; in the USA with the influx of Hispanics; and in Australia where there have been successive waves of migrants from all parts of the world since 1945. In Australia the problem looms large for three reasons. The population of non-English speaking background is ageing more rapidly than the general population. Family reunion migration of the aged, especially in recent years from Asia, is a further factor. Thirdly, waves of immigration to Australia will result in successive cohorts of Eastern European, Mediterranean, Arabic, and South East Asian migrants reaching the senium well into the next century without the development of matching specialized ethnic services. This is clear from Australian Bureau of Statistics projections for the periods of peak demand for migrants aged 75 or more. From Poland this period will be between 2001–6, from Italy 2006–11, from Greece 2011–16, and from Yugoslavia, India, Malta, and China 2016–21 (quoted in Rowland 1991, p.139). Just as the number of frail aged from different origins is destined to reach a maximum at different times, so too will the height and duration of these peaks vary as a result of past trends in immigration (Rowland 1991).

It is unlikely that mainstream services will in the near future have the resources to meet these demands. Short-term strategies required to cater for these needs include: interpreters, ethno-specific services, the use of ethnic community voluntary groups, and recruitment and training of appropriate personnel from within ethnic communities. In the long-term, better training for mainstream workers is required. One creative scheme currently underway in nursing homes is clustering. This refers to the aggregation of residents of similar background and/or language in the one facility thereby obviating the need to establish separate nursing homes for each ethnic group. This minimizes costs but creates obvious advantages for residents in the local community. In day centres it is possible to dedicate individual days to specific ethnic or language groups and once again maximize resources without increasing costs.

Younger persons with dementia

The family of a 50-year-old person with dementia at present fully employed and with teenage children faces additional strains. Adolescents negotiating their own developmental stages seem to react particularly strongly to their parent's illness; rejection, anger, and problems with school are common. The theme of double-isolation recurs in this situation as the family is confronted by services geared to a much older clientele. This is particularly apparent for the young male sufferer when admitted to a nursing home where the average age is over 80 years and females far outnumber men.

Non-degenerative causes of dementia are more likely in younger persons and may be associated with specific problems for families. Examples include the fear of infection and dementia resulting from AIDS or Creutzfeldt-Jakob disease, the dread of inheriting Huntington's chorea, and the stigma of syphilis. In general, a younger age of dementia onset confers an increased risk of heritability to children and siblings of probands. This pattern occurs in Alzheimer's disease where the heritability risk, being inversely proportional to age of onset in the proband, can approach 50 per cent.

Professionals and VIPs

Doctors, lawyers, architects, many other professionals, politicians, and the very wealthy are often isolated by their special status and the awkwardness others feel in their clinical care. Assessments are delayed for fear of jeopardizing careers and this may place others in danger and the family in a quandary. Perceived class barriers to care and community services may further increase family burden.

Second marriages and alternative relationships

When one partner starts to decline only a few years after remarriage, the other, unbound by a lifelong debt of mutual assistance, may feel cheated, not sustain affection in the face of dementia, and more quickly arrange residential care. Often the children from each spouse's first marriage have recriminations: those of the ill spouse accusing the step-parent of exaggerating deficits; and those of the well spouse resenting their parent's new-found burden and accusing their step-siblings of a lack of help. Childless marriages where there is less support available, homosexual partners coping with the dementia of AIDS, and couples where both are dementing, also face special problems.

Families who abuse older people

Elder abuse is an increasingly recognized occurrence affecting 3.2 per cent of older people in the USA (Pillemer and Finkelhor 1988). Abuse may be physical, sexual, psychological, financial, or by neglect. Typically the abuser is a family member, a spouse (50 per cent of the time) or a child or grandchild (40 per cent of the time) (Bourland 1990). The abuser may have grown up in a culture of family violence and repeats this cycle – the abused child growing up and abusing his own children and elderly parents. The abuser may have a history of drug or alcohol use or of psychiatric disorder. Often, of several children, the one least socially integrated, competent and able to provide care, for example, an unemployed single daughter inept at making relationships and unskilled in any occupation, is called upon to move in with the dementing parent as she is considered to be most available.

The abused person is most often a female with cognitive impairment, and usually financially and physically dependent. The co-existence of psychosis or problem behaviours such as aggression or constant questioning on the part of the demented person increases the likelihood of elder abuse.

Intervention must be tailored to the situation (Bourland 1990). Where the relative or spouse carer is violent, psychopathic, or alcoholic it is usually prudent to separate the older person from the abuser, often under the guise of admission for a medical diagnosis. The older abused person is usually apprehensive of change and may defend the accuser against accusations out of fear of reprisal. It may be necessary to instigate Guardianship or other legal mechanisms in order to effect alternative arrangements.

On the other hand, where abuse occurs in the setting of an exasperated carer with limited ability to cope, intervention is shaped around relieving carer stress by arranging support services, respite care, and supportive counselling for the family.

HELPING FAMILIES CARE FOR THEIR RELATIVE WITH DEMENTIA

Helping by stages

As discussed earlier, dementia has a career and in the different stages of its progression different strategies are required. In the early stages, counselling of the dementing person and the family about the diagnosis, management, and prognosis are required. Financial and legal planning have been discussed earlier. Depression, anticipatory grief, and anxiety about the future are frequent foci for supportive and expressive therapy encouraging the person to ventilate troubled emotions. Planning about work, family, and the future in general should be considered. Discussion with the extended family, and other people of significance can be useful in mobilizing help for the principal carer. Sessions with the extended family may provide a flashpoint, igniting longstanding family feuds; or may douse long-smouldering fires, altering patterns of family interaction towards co-operativeness and drawing families and friends together to provide care as best they can.

During the middle stages, community services, medical and specialist consultations to assist with problem behaviours, and support for the immediate family are required. Psychiatric complications of dementia such as depression, hallucinations, delusions, and misidentifications are common. Often the family is more concerned about the abnormal behaviour or the hallucinations than is the person with dementia. In such case, the intervention is aimed at the family.

Towards the late stage greater nursing and professional services are required. Carers find it difficult to judge the critical point at which they should arrange nursing home admission. It may be determined by excessive dependence (for example, double incontinence), the necessity for 24-hour care (for example, severely disrupted sleep pattern), the dementing person's inability to recognize the carer or illness in the carer.

Financial implications for nursing home admission vary between countries. In Australia and the UK there is little financial disincentive for admission but in the US families face major costs and the phenomena of 'granny dumping' and couples divorcing to become eligible for Medicaid have been reported.

Post-bereavement

Suddenly the funeral without an end is at an end. The carer, whose interests have narrowed and whose hobbies have been discarded, is alone. A life dominated by dementia and daily visits to the nursing home – the centre-piece of her daily routine – is no more. The companion who was both absent and present at least provided a *raison d'être*. Well-meaning comments from friends about death being a welcome release are nobly but uneasily accepted. Unfortunately it is a time when professional services cease their involvement too.

Support services

There is evidence (see above) that informal supports are more helpful to families than formal, professional services. While the effects of the latter are difficult to assess, there is no doubt that many families clamour for more professional help, particularly respite care. Superimposed on their own desire to provide care and their feelings of guilt, societal expectations and financial disincentives may lead families to delay arranging nursing home admission. As governments increasingly strive to maintain dementing people in the community for longer more help will be required from families, friends, and professionals.

Formal support services include domiciliary nursing, Meals-on-Wheels, Day Centres, and respite care. There are several types of respite care. Residential respite care can be planned, regular, and prophylactic, for example, two weeks every three to six months to prevent care-giver breakdown or emergency, for example, at time of care-giver illness. In-home respite provides relief for the care-giver by having a paid person coming into the dementing person's home, for a day, overnight, or longer.

Self-help groups

Concurrently with the world-wide recognition of Alzheimer's disease and other dementias has been the rise of Alzheimer's self-help associations. These Associations provide regional self-help groups, help break down social isolation, and offer practical advice and emotional support. For some, the self-help group provides carers with an identity; for others, a second family to replace the one currently disintegrating. Additionally Alzheimer's associations lobby governments and bureaucracies for better conditions for sufferers, more services and financial support for families, and more research into causes and treatments of the dementias.

Intervention programmes

The basic premise of most programmes is that the course of dementia can be altered and/or the effects on carers ameliorated by intervening with carers. There has been a growing number of studies since the 1980s, and virtually none before that, describing or evaluating intervention programmes with carers.

Elements of programmes

Psychological: Most programmes provide some support usually allowing for ventilation and the development of group process (Simank and Strickland 1986). They enable participants to share ideas, empathize with each other, and acknowledge and affirm each other's care-giving efforts. Participants learn how to express their feelings, for example, guilt, anger, and loneliness, and to cope better with their highly emotional task, for example, decision to institutionalize (Toseland and Rossiter 1989).

Education: a feature of many programs (for example, Zarit *et al.* 1987; Brodaty and Gresham 1989) usually deals with the medical, psychological, social, financial, and legal aspects of dementia. Programmes may teach specific skills and problem-solving approaches to carers: how to provide mental stimulation for a person with dementia, how to engage him therapeutically in activities, the use of reminiscence, motivational enhancement, assistance with basic daily living skills such as lifting and bathing, the use of prosthetic appliances and devices, and administration of medications. Other topics include home modification to facilitate care-giving, handling financial matters, and dealing with legal issues (see, Brodaty 1992).

Outcome of intervention

The few consistent results emerging from outcome studies of these programmes suggest that they improve carer knowledge, decrease family burden, improve coping skills, and are rated by participants as helpful. Some studies have found that carers' depression scores decrease after intervention. No study has yet shown improvement in carer social functioning.

Importantly, family carer training programmes may be cost-effective (Brodaty and Peters 1991) and delay institutionalization (Brodaty and Gresham 1989; Greene and Monahan 1987). Parallels can be drawn with programmes which have beneficial effects for families of those suffering from other chronic psychiatric disorders such as schizophrenia. It is unlikely that carer training is suitable for all. Undoubtedly the skill is to match the family, the illness, the problems, and the training techniques.

CONCLUSIONS

Dementia is a progressive debilitating condition that has major negative effects on families and society. Family carers, particularly women, bear the brunt of the burden for many years, their distress levels are high and some are unable to manage. Results from research have enabled carers vulnerable to depression or breakdown to be identified. The strain of dementia on carers can be ameliorated by effective coping styles, family support, community services, and carer training programs. Empirical evidence for the utility of carer interventions is however mixed, suggesting a need for specificity in matching intervention to the illness, the stage of the dementia, the individual, and the carer.

REFERENCES

Anderson, R. (1987). The unremitting burden on carers. *British Medical Journal*, **294**, 73–4.

Aronson, M.K., Ooi, W.L., Morgenstern, H. *et al.* (1990). Women, myocardial infarction, and dementia in the very old. *Neurology*, **40**, 1102–6.

Beck, A.T., Ward, C.H., Mendelson, M. *et al.* (1961). An inventory for measuring depression. *Archives of General Psychiatry*, **4**, 561–71.

Bledin, K.D., MacCarthy, B., Kuipers, L., and Woods, R.T. (1990). Daughters of people with dementia. Expressed emotion, strain and coping. *British Journal of Psychiatry*, **157**, 221–7.

Blessed, G., Tomlinson, B.E., and Roth, M. (1968). The association between quantitative measures of dementia and of senile changes in the cerebral grey matter of elderly subjects. *British Journal of Psychiatry*, **114**, 797–811.

Boss, P., Caron, W., Horbal, J., and Mortimer, J. (1990). Predictors of depression in caregivers of dementia patients: boundary ambiguity and mastery. *Family Process*, **29**, 245–54.

Bourland, M.D. (1990). Elder abuse: from definition to prevention. *Postgraduate Medicine*, **87**, 139–44.

Brodaty, H. (1992). Carers: training informal carers. In *Recent advances in psychogeriatrics 2* (ed. T. Arie). Churchill Livingstone, London.

Brodaty, H. and Gresham, M. (1989). Effect of a training programme to reduce stress in carers of patients with dementia. *British Medical Journal*, **299**, 1375–9.

Brodaty, H. and Hadzi-Pavlovic, D. (1990). The psychosocial effects on carers on living with dementia. *Australian and New Zealand Journal of Psychiatry*, **24**, 351–61.

Brodaty, H. and Peters, K.E. (1991). Cost effectiveness of a training program for dementia carers. *International Psychogeriatrics*, **3**, 11–22.

Brody, E.M. (1981). 'Women in the middle' and family help to older people. *The Gerontologist*, **21**, 471–80.

Brody, E.M., Kleban, M.H., Johnsen, P.T., Hoffman, C., and Schoonover, C.B. (1987). Work status and parent care: a comparison of four groups of women. *The Gerontologist*, **27**, 201–8.

Challis, D., Chessum, R., Chesterman, J. *et al.* (1987). Community care for the frail elderly: an urban experiment. *British Journal of Social Work*, **18**, 13–42.

Chenoweth, B. and Spencer, B. (1986). Dementia: the experience of family caregivers. *The Gerontologist*, **26**, 267–72.

Chesterman, J., Challis, D. and Davies, B. (1987). Long-term care at home for the elderly: a four-year follow-up. *British Journal of Social Work*, **18**, 43–53.

Cohen, D. and Eisdorfer, C. (1988). Depression in family members caring for a relative with Alzheimer's disease. *Journal of the American Geriatrics Society*, **36**, 885–9.

Colerick, E.J. and George, L.K. (1986). Predictors of institutionalization among caregivers of patients with Alzheimer's disease. *Journal of the American Geriatrics Society*, **34**, 493–8.

Draper, B., Poulos, C., Cole, A. *et al.* (1992). A comparison of caregivers of elderly stroke and dementia victims. *Journal of the American Geriatrics Society*, **40**, 886–991.

Edwards, J.R. and Cooper, C.L. (1988). Research in stress, coping and health: theoretical and methodological issues. *Psychological Medicine*, **18**, 15–20.

Eisdorfer, C. (1991). Caregiving: an emerging risk factor for emotional and physical pathology. *Bulletin of the Menninger Clinic*, **55**, 238–47.

Enright, R.B. (1991). Time spent caring and help received by spouses and adult children of brain-impaired adults. *The Gerontologist*, **31**, 375–83.

Fiore, J., Coppel, D.B., Becker, J., and Cox, G.B. (1986). Social support as a multifaceted concept: examination of important dimensions for adjustment. *American Journal of Community Psychology*, **14**, 93–111.

Fitting, M., Rabins, P., Lucas, J., and Eastham, J. (1986). Caregivers for dementia patients: a comparison of husbands and wives. *The Gerontologist*, **26**, 248–52

Gilleard, C.J. (1984). *Living with dementia*. Croom Helm, London.

Gilleard, C.J. (1992). Carers: recent research findings. In *Recent advances in psychogeriatrics 2* (ed. T. Arie). Churchill Livingstone, London.

Gilleard, C.J., Boyd, W.D., and Watt, G. (1982). Problems in caring for the elderly mentally infirm at home. *Archives of Gerontology and Geriatrics*, **1**, 151–8.

Goldberg, D.P. (1972). *The detection of psychiatric illness by questionnaire*. Oxford University Press, London.

Greene, V.L. and Monahan, D.J. (1987). The effect on a professionally guided caregiver support and education group on institutionalization of care receivers. *The Gerontologist*, **27**, 716–21.

Greene, J.C., Smith, R., Gardiner, M., and Timbury, G.C. (1982). Measuring behavioural disturbance of elderly demented patients in the community and its effects on relatives: a factor analytic study. *Age and Ageing*, **11**, 121–6.

Hall, J.N. (1990). A psychology of caring. *British Journal of Clinical Psychology*, **29**, 129–44.

Jorm, A.F. and Henderson, A.S. (1990). Department of Community Services and Health *The problems of dementia in Australia (2nd edn)*. Australian Government Publishing service, Canberra.

Kiecolt-Glaser, J.K., Glaser, R., Shuttleworth, E.C. *et al.* (1987). Chronic stress and immunity in family caregivers of Alzheimer's disease victims. *Psychosomatic Medicine*, **49**, 523–35.

Lawton, M.P., Moss, M., Kleban, M.H. *et al.* (1991). A two-factor model of caregiving appraisal and psychological well-being. *Journal of Gerontology*, **46**, 181–9.

Lazarus, R.S. and Folkman, S. (1984). *Stress, coping and adaptation*. Springer, New York.

Lezak, M.D. (1978). Living with the characterologically altered brain injured patient. *Journal of Clinical Psychiatry*, **39**, 592–8.

Mace, N.L. and Rabins, P.V. (1981). *The 36-hour day*. Johns Hopkins University Press, Baltimore.

Morris, L.W., Morris, R.G., and Britton, P.G. (1988). The relationship between marital intimacy, perceived strain and depression in spouse caregivers of dementia suffers. *British Journal of Medical Psychology*, **61**, 231–6.

Pillemer, K. and Finkelhor, D. (1988). The prevalence of elder abuse: a random sample survey. *The Gerontologist*, **28**, 51–7.

Porter, B.E. and Robinson, M.B. (1987). *Protected persons and their property in New South Wales*. Law Book Company, Sydney.

Poulshock, S.W. and Deimling, G.T. (1984). Families caring for elders in residence: issues in the measurement of burden. *Journal of Gerontology*, **39**, 230–9.

Pruchno, R.A. and Resch, N.L. (1989). Husbands and wives as caregivers: antecedents of depression and burden. *The Gerontologist*, **29**, 159–65.

Roth, M. (1955). The natural history of mental disorders in old age. *Journal of Mental Science*, **101**, 281–301.

Rowland, D.T. (1991). Care of the ethnic aged. *In Aged care reform strategy mid-term review 1990–91, discussion papers*. Australian Government Publishing Services, Canberra.

Schulz, R., Vistainer, P., and Williamson, G.M. (1990). Psychiatric and physical morbidity effects on caregiving. *Journal of Gerontology*, **45**, 181–91.

Sherrin, C.H., Barlow, R.F.D., and Wallington, R.A. (1987). *Williams' law relating to wills* (6th edn), Vol. 1. Butterworths, London.

Simank, M.H. and Strickland, K.J. (1986). Assisting families in coping with Alzheimer's disease and other related dementias with the establishment of a mutual support group. *Journal of Gerontological Social Work*, **9**, 49–58.

Spar, J.E. and Garb, A.S. (1992). Assessing competency to make a will. *American Journal of Psychiatry*, **149**, 169–74.

Stone, R., Cafferata, G.L., and Sangl, J. (1987). Caregivers of the frail elderly: a national profile. *The Gerontologist,* **27**, 616–26.

Toseland, R.W. and Rossiter, C.M. (1989). Group interventions to support family caregivers: a review and analysis. *The Gerontologist*, **29**, 438–48.

U.S. Congress Office of Technology Assessment. (1987). *Losing a million minds, confronting the tragedy of Alzheimer's disease and other dementias. U.S. Government Printing Office, Washington, DC.*

Wells, Y.D., Jorm, A.F. Jordan, F., and Lefroy, R. (1990). Effects on care-givers of special daycare programmes for dementia suffers. *Australian and New Zealand Journal of Psychiatry*, **24**, 82–90.

Wright, L.K. (1991). The impact of Alzheimer's disease on the marital relationship. *The Gerontologist*, **31**, 224–37.

Zarit, S.H. and Toseland, R.W. (1989). Current and future directions in family caregiving research. *The Gerontologist*, **29**, 481–3.

Zarit, S.H., Anthony, C.R., and Boutselis, M. (1987). Interventions with caregivers of dementia patients: comparison of two approaches. *Psychology and Aging*, **2**, 225–32.

13

Family violence

Terms such as abuse or violence describe a wide range of behaviours, which is a problem for social scientists, clinicians, police, and legislators, who attempt to describe the extent and seriousness of the problem and to find the most effective means of reducing its prevalence. Criminal, sociological, behavioural, psychiatric, and medical approaches differ widely in the concepts they use, their methods of data collection, and the solutions they propose. Most research of the non-anecdotal kind has concentrated on the most severe forms of physical acts of violence which are more easily defined and whose immediate physical effects are more readily described and measured than are the psychological and less severe physical forms of injury and intimidation. Family violence includes:

1. Spouse abuse (most often violence against women). This includes verbal insults, threats of physical violence, coercive sex (including marital rape), and physical injury which may be of life-threatening severity.

2. Child abuse – including physical and sexual abuse (incest), child neglect (physical and emotional), and infanticide.

3. Sibling violence.

4. Violence by children towards parents – including matricide and patricide, emotional abuse, social and economic exploitation, all of which have been described in the increasingly prevalent phenomenon of elder abuse.

Although family violence is not a unitary phenomenon (Frude 1991), there are strong associations between the different forms of violence, and indeed these may coexist in the one family. A person therefore may be the victim of several forms of violence, as well as of the family environment which permitted or facilitated this violence to occur. It is extremely difficult to assess the comparative influence of each of these factors in studying the long-term sequelae of family violence on its victims.

Research, clinical intervention, and social reforms in dealing with this problem have been hampered by societal attitudes toward the 'sanctity' of the family, which have led the family and friends of the victim, doctors, judges, and police to minimize the seriousness of the problem in an individual case or society at large.

Feminists argue that such attitudes reflect the traditional patriarchal ideology where women and children are perceived as men's property, over whom they have a natural entitlement (see Chapter 14); furthermore the family is romanticized and idealized as a 'sanctuary' where men may exert their will without interference from the state. Some feminists also argue that psychiatric formulations of victims (that is, survivors) and perpetrators, or the typologies and dynamics of 'dysfunctional' families where violence occurs, may distract from the urgent need to reform the social institution of the family via educational, legal, and legislative means.

However, it is our view that explanations of behaviour do not count as justifications, and that the domains of knowledge and social justice should inform rather than undermine one another. The practical implications of this view are examined by Bograd (1992).

SPOUSE ABUSE

Physical violence of a severity that does not result in death of the victim is committed by men against women many times more frequently than by women against men. The exact figure in any given community is hard to ascertain because of under-recording, but the male-to-female ratio of physical abusers ranges from approximately 10:1 to 100:1. Even when the violence is committed by both partners against each other (about 4 per cent of recorded incidents are of 'mutual violence'), the consequences for the woman are usually far more serious than for the man in terms of severity of injury and her ability to escape from the violent relationship. About 50 per cent of female victims of homicide are murdered by their husband, boyfriend, or male cohabitee, but only 10 per cent of male homicide victims are killed by a wife or girlfriend. The majority of women who kill a man do so in self defence, or as a desperate attempt to avoid further physical and/or sexual abuse against themselves or their children; this is rarely the case with male murderers (Pagelow 1984).

Wife beating of varying severity has been recorded throughout history, in all social, economic, ethnic, and religious groups. Psychoanalytic anthropologists suggest that the devaluation of and the violence against women are culturally ingrained patterns, which reflect men's deep anxieties about their dependency on women. These scholars point to the myths and social rituals about menstruation, child-bearing, witchcraft, and other supposedly destructive influences that women have on men which are found in all cultures.

For clinical purposes, men's violence against women in the home has been studied in terms of three sets of interrelated factors, that is, social stressors, marital interaction, and individual factors in both the perpetrator and the victim. A 'diathesis-stress' model proposes a continuum of vulnerability or propensity for individuals to commit acts of violence against other family members based on an interaction of these three factors.

Environmental factors

Unemployment, poverty, and crowded living conditions are associated with increased risk of violence in families. This may be due to the interplay of a number of factors, including the increased time and physical proximity that the partners have to one another under such conditions, personality factors that may predispose to create these environmental conditions, and the effects of such conditions on self-esteem. Social isolation is also linked with an increased risk of family violence. Again this may reflect lack of social skills and poor self-esteem. Unemployment and poverty too may increase the degree of social isolation, as may the violence itself.

Alcohol is often used to relieve the tension and boredom that difficult environmental conditions impose, and may lower the threshold of violence (see below).

While lower socio-economic status is widely linked to family violence three cautionary points must be made before a causal influence is invoked. Firstly, it is more likely that violence in families from lower socio-economic groups will be reported to the authorities. Furthermore, it is possible that the helping professions, most of whose members are middle-class, will take the report of violence more seriously if the family is from a lower socio-economic class. Thirdly, alternative means and opportunities for dealing with various stressors may be less available to lower income households. These considerations also apply to reports of increased family violence among racial minority groups in Western societies.

Further environmental stressors may be associated with life-cycle changes such as an unwanted pregnancy, the birth of a handicapped child, death of significant family members, migration, a child leaving home, or recurrent illness in children or other family members.

Marital relationship factors

The risk of violence is increased in marriages characterized as unstable or disorganized, or where frequent separations or threats of separation occur (Frude 1991). The marital instability may be due to a previous history of violence in the marriage and leads to further violence in circumstances where the precipitant is that of relatively minor conflict. The risk of violence increases over time, as does its intensity and frequency. This may reflect the increasing instability of the marriage and the risk of separation. Indeed, the chance of violence increases in the wake of the wife's decision to leave a marriage, or after the actual separation or the legal formalization of the divorce (Pagelow 1984).

The pattern of violence may be established before the couple cohabit or marry, reflected in the phenomenon of 'dating' violence. Most studies conducted on samples of USA college students show an incidence from 20 per cent to 60 per cent. Over half the women who reported dating violence continued to date their partner. The violence begins after the relationship has become sexual or when the couple regard their relationship as 'serious' (Cate *et al.* 1982)

A similar pattern of 'dating' violence has been noted among high-school students. In over one-third the relationship continued after the violence, the protagonists viewing it as an expression of love. This perverse interpretation of violence indicates a need for children to be educated about the differences between love and domination.

Inequalities in the power relationship between men and women are part of the cultural norms in many societies. Where the man has a low social status in absolute terms, or where he has a lower social status than his wife the risk of violence in the marriage is increased.

Individual factors

The violent man

Biological aspects are often overlooked in many accounts of family violence but may be significant, and should be assessed. A history of premature or difficult birth, childhood learning disabilities, head trauma (including repeated physical violence during childhood), epilepsy or motor vehicle accident, all suggest a possible neurological deficit.

Alcohol consumption and its effects are highly relevant here. The abrupt onset of violent behaviour in a man previously non-aggressive, especially if coupled with other changes in personality, suggests an underlying brain illness or a primary psychiatric disorder, especially affective illness.

Psychological – over half of repeatedly violent men have witnessed recurrent episodes of physical violence by their fathers or other adult males against their mother; about a third of physically abusive men were themselves childhood victims of physical abuse, usually by the father or an adult male in the home. This has given rise to the popular concept of the 'cycle of violence' whereby violent or sexually abusive behaviour is transmitted across the generations. However, this fatalistic concept has been challenged and will be discussed later.

The psychological features of the typically violent man include low self-esteem, unmet dependency needs, difficulty in appropriate assertive-ness, mistrust of emotions (especially those considered to be feminine), a tendency to substitute action for feeling, and the over-valuation of self-reliance which belies his fragile sense of self.

The perpetrator marries a woman whom he perceives as embodying traditional feminine attributes which will serve to meet his needs for nurturance, and the efficient care of the family. Children are perceived as duty-bound to be respectful, loving, and obedient to their father.

Jealousy and exaggerated protectiveness of wife and children (especially daughters) are common, and in its more severe forms pathological (morbid) jealousy may dominate. At other times the man's sense of grievance against the world and feelings of entitlement to receive respect from all are directed against wife and children. He will also feel betrayed by them should they disclose the violence.

Through his wife and children the perpetrator of violence seeks to solve his sense of personal deficiency by creating an idealized family. If these idealized relationships cannot be sustained or when challenged by stress, he becomes frustrated and may express this through violence. Dealing with anger in this way may be socially learned in the man's family of origin, as well as being culturally sanctioned. It is not often realized that many of these men are fearful of their own (or other's) anger, precisely because they know of no way of expressing it other than by violence. This results in a degree of denial so that when ques-tioned about violence he will minimize its severity and attribute it to external factors beyond his control or as a response to provocation by the victim.

As will be discussed later, similar characteristics are observed in adults who assault or abuse their children physically, sexually, or emotionally.

The woman victim
No typical profile of the victim can be drawn. Women who come to the attention of the police or helping agencies may not represent those who either continue to live in or leave such a relationship.

Family members, friends, and professional helpers are often curious about why a woman apparently chooses to remain for so long in a relationship with a violent man. While no single causal factor can be identified common reasons include (Pagelow 1984, pp. 306–13):

1. Fear of further violence, including possible harm to the children. A woman who decides to leave may be impeded by him and the children may be victimized. She may fear for the safety of family and friends who offer support and shelter, or fear further violence at the inevitable meetings she will have after a separation, for example, court hearings and during access visits to the children.

2. Financial disadvantages incurred by being a single parent, lack of alternative housing, job opportunities, and reduced educational and material support for the children (since it is usually the mother who is the custodial parent). Such burdens are aggravated if a child is chronically ill or handicapped.

3. The perceived needs of the children for an intact home life and for a relationship with their father, despite his violence. The woman may have been socialized to place her children's needs ahead of her own; in addition she may also try to give her children a relationship with their father that she was denied in her childhood.

4. Expectations of the woman's family and friends that she remain with her husband. They may disbelieve her descriptions of violent behaviour and their incredulity may also be shared by the police and the helping professions.

5. Socialization of the woman in patriarchal society emphasizes her role as homemaker. Her 'failure' to fulfil this arouses guilt, shame, and sense of unworthiness, which may amplify pre-existing feelings from her own childhood relationships with violent or neglectful parents.

6. The woman may blame herself for her husband's violence. Usually domestic disagreements over sexual, financial, or other matters serve as a catalyst, the woman concluding that had she been more conciliatory the violence may not have occurred.

7. The promise of change. After an episode of violence the man is often contrite, apologetic, loving, and promises to change. He may enlist children and other family members or friends to 'hear his confession' and his new resolve. The wife may feel needed, and because of her own emotional experiences she identifies with the distress of the husband and her children, and may experience the period of reconciliation as loving and deeply satisfying. Couples who share a borderline personality organization are particularly prone to this kind of behaviour (Gillman 1980).

8. Concern for the spouse's welfare. The violent man is often psychologically immature, unstable, and apt to act out his distress by resort to alcohol, attempted suicide, reckless driving, brawling, and other self-destructive behaviours. The wife may fear the effect of her leaving will cause him to 'break down'.

9. Psychic numbing and learned helplessness. The woman who, for years was subjected to physical and-or psychological abuse may develop a mental

state of emotional numbness, passiveness, and learned helplessness. Use of alcohol or drugs may add to her mental inertia. Although this may have had survival value when she was trapped in the oppressive domestic environment, it limits her capacity to face all the risks and obstacles in a decision to leave.

10. The intermittent nature of the violence especially if it is not severe in physical terms, combined with the behaviour of the perpetrator between violent episodes when he is affectionate and reliable, may further confuse the woman's assessment of the seriousness of her situation.

This daunting list of problems illustrates the enormity of the task faced by many women who seek to leave their violent husbands. In the light of these difficulties we can understand the concern of feminist therapists that by considering family systems approaches and family dynamics the urgent need for protection and assistance for women and children victims is obscured and the perpetrator is permitted to avoid responsibility for his violent actions (Bograd 1984).

However, some clinically useful formulations may be made. We have already referred to the marital dynamics which may sustain alternating periods of rage-laden violence and loving reconciliation.

Transgenerational patterns of family interaction have also been described in which the victim and perpetrator repeat or seek to heal relationships with their own violent, neglectful, or idealized parents.

Studies of women who leave a violent relationship and seek refuge in a women's shelter indicate that almost half return to the relationship, and that in over 50 per cent of these further violence occurs within 6 months (Frude 1991). Whether the woman's behaviour reflects the huge practical obstacles and social pressures she faces in attempting to separate from her husband, or the influence of the transgenerational and marital dynamics is unclear. It is likely that both sets of factors operate in varying measure.

Effects of violence

The long-term effects of domestic violence on *women* include stress-related physical illness, chronic depression and/or anxiety, dissociative states, alcoholism, substance abuse, and attempted suicide. These effects are aggravated if the children also have emotional or behavioural difficulties. Sometimes women may also physically abuse or neglect their children.

It is difficult to study the direct effect on *children* of their having witnessed repeated physical violence between parents, since its consequences on parents and their relationship will also influence the children's response. Furthermore, a proportion of children will have been physically abused or intimidated or neglected by one or either parent (see below).

Children who witness repeated physical violence in the home are likely to display a range of behavioural problems including inappropriate aggression or excessive passiveness, chronic anxiety, depression, phobias, excessive crying and clinging. School-age children may feel guilty about their inability to protect their mother, while younger children may worry that they are responsible for the violence. Gender differences also appear relevant, with pre-school boys and school-age girls being most affected. Not surprisingly, children who have experienced physical abuse as well as witnessing violence between their parents have increased rates of psychopathology compared with children who only witnessed parental violence (Hughes 1988). The multiple traumas, losses, and loyalty conflicts experienced by children whose mother is killed by their father and principles of therapeutic management have been described (Black and Kaplan 1988).

Over half the boys who witness repeated physical violence during childhood go on to violent relationships in their adult years and accept violence as an appropriate way to resolve disputes to a greater degree than children who have not witnessed repeated violence. Sophisticated methodology is needed to investigate the factors which lead almost half the boys and over two-thirds of girls who have witnessed family violence to avoid repeating this behaviour in their own marital and parenting life.

Coercive sexual intercourse (marital rape)

A form of men's violence towards women that has generated social interest and judicial controversy is coercive sexual intercourse in a marriage or an established relationship. This problem represents the intersection of two concerns raised by the women's movement, family violence and rape, both of which have been tolerated and even encouraged under patriarchal forms of domination.

Three patterns of marital rape have been described (Yllo and Strauss 1981), although this typology awaits further investigation: (a) as part of a violent relationship in which sex is used to control and intimidate the woman; (b) in an otherwise non-violent relationship, in which the husband attempts to gain control over the frequency or type of intercourse in order to satisfy his needs; and (c) 'obsessive' rape, where the man's severe character pathology is reflected in sexually sadistic behaviour and interest in pornography, including child sexual pornography.

Women vary in their willingness to report forced marital intercourse, and their views of this as rape. This is not surprising since the legal category of marital rape is still in question and many women have been socialized to believe that they are not entitled to refuse their husband's demand for sex.

Coercive sex can be highly traumatic (Russell 1982). In addition to physical harm it may result in considerable emotional distress, which may reach the intensity of a post-traumatic stress reaction, similar to that observed in rape. Not surprisingly, women subjected to both sexual and non-sexual violence tend to have more severe and chronic reactions than do the victims of non-sexual marital violence. The emotional consequences are aggravated further by the likelihood that coercive sex in marriage is likely to be repeated many times over a period of many years.

A clinician should enquire specifically about such behaviour in any case of marital difficulties, divorce, sexual dysfunction, or where the woman presents suffering from 'psychosomatic' problems, alcoholism, substance abuse, or attempted suicide.

A greatly neglected issue is the physical and sexual abuse of mentally ill and intellectually handicapped people, men as well as women and children, though women may be at particular risk from their male caregivers in the home and in institutions (Carlile 1991).

Treatment of wife abuse

A range of approaches exists, many based on myths and misunderstandings and few have been adequately evaluated (Dickstein 1988). An emerging consensus recognizes the need to protect the victim from further violence but acknowledges that this is unlikely to be effective unless the perpetrator accepts responsibility and is willing to change.

The first aim of therapy is to stop the violent or abusive behaviour rather than to break up the family. It consists of: (a) the separation of the victim from the perpetrator when the possibility of further violence exists; (b) separate counselling for perpetrator and victim, individually or in groups: additional treatment for specific problems such as alcohol and drug abuse may also be required; (c) marital or family counselling when the couple are not living together the same household; (d) marital or family counselling when the perpetrator returns to live with the family.

Therapy for the woman alone or immediate marital or family therapy is inappropriate if they encourage the perpetrator to believe that the marriage can continue without his taking responsibility for ceasing the violence.

Therapy of the victim

The single most effective intervention to decrease marital violence is police willingness to lay criminal charges against the perpetrator (Sherman and Berk 1984). Such action is more effective than their giving of advice, warning or separating the couple.

An alternative is the imposition of a legal restraining order until the court is satisfied that the perpetrator has shown evidence of containing his violence after a period of mandatory therapy (Jenkins 1991).

Many communities have established hot-lines whereby women can contact police in an emergency, and many police forces have undergone special training for intervening in cases of domestic violence.

A further societal response is the growing number of *refuges* or shelters provided by the State or privately funded organizations where a woman and her children may find temporary accommodation in order to escape a violent man. In order to protect the shelters and their occupants from the possible violence of aggrieved men knowledge of their location is confined to workers and social agencies who deal specifically with this problem.

The refuges also provide a forum for the woman to review her marital relationship, to make contact with family, friends, other women's support groups, and various social agencies which may help to overcome the sense of isolation, helplessness, and fear that the violent relationship has created. Since many women return to live with their violent partner the period in the shelter is also used to lay the groundwork for couple therapy.

Hospitalization is considered if a shelter or an appropriate alternative is unavailable and there is a risk of further injury if she returns home. Hospitalization should also be considered for a woman suffering from acute post-traumatic stress disorder, dissociative state, severe anxiety, acute depression, or where the risk of suicide or attempted suicide is appreciable. A working relationship between hospital and the refuge staff is important for devising management. Extended family and friends may be involved, especially where children require care.

It is important that helpers not hurry the woman toward separation or independence ahead of her capacity to deal with feelings about past trauma and future changes. Walker (1979) summarizes the contradictory experiences which the woman must resolve before constructive, realistic planning is possible. These include love and hate toward the perpetrator, rage towards and fear of the perpetrator, wishes to remain and to leave the relationship, feeling secure, and the panic of abandonment.

The aim of *psychotherapy* is to help the woman decide whether to end the violence or the relationship. This may be conducted on an individual and/or group basis. Core issues are impaired self-esteem, emotional isolation, and mistrust (Hilberman 1980). Feelings of guilt, rage, and fear of losing control occur and she may find it hard to reconcile these feelings with her own sense of victimization. The feelings are even more difficult to resolve when the woman has herself been physically or emotionally abusive to her own children. Furthermore, her experience in her family of origin and in society have taught her to suppress her anger, unlike men (for example, her father, brothers, lovers, and husband) who are permitted expression of such feeling.

Despite this double standard, self-assertiveness training may be hazardous for the woman if her husband is not learning appropriate responses to her attempts to express herself more assertively. Similarly, a clinical approach which adopts a naive systems view that because both partners cue each other reciprocally in a way which culminates in violence, they are equally responsible, is now regarded as exonerating the husband and adding to the woman's already considerable burden. A growing number of individual and family psychotherapy approaches have modified their traditional methods in the light of such criticism, with feminist-informed therapy a feature of psychodynamic (Gillman 1980), existential (Weingourt 1985), and systems (Goldner *et al.* 1990) approaches. However, the therapist who criticises the perpetrator and exhorts the wife to leave without appraising the significance of the relationship for the woman, risks turning therapy into a power struggle (Stulberg 1989).

The therapy of the perpetrator

Comparatively little clinical or research attention has been paid to the treatment of perpetrators, most of whom do not seek or fail to attend therapy. The extant literature is impressionistic, usually derived from a clinical experience with a small number of offenders, and while reported therapy appears to have been conducted with skill, it is difficult to extrapolate to the problem of family violence in general. Long-term follow-up studies are almost completely lacking.

One feature commonly reported is the perpetrator's unwillingness or inability to acknowledge that his violent behaviour is inappropriate or to accept responsibility (Koval *et al.* 1982; Jenkins 1990). Psychiatrists and other clinicians who have been trained to be non-judgemental about their patients' behaviour may reinforce this attitude, as may exploration of the underlying personality conflicts or deficits presumed to cause violent behaviour.

By contrast, the 'invitation to responsibility' approach (Jenkins 1990) uses a model based on a style of questioning that encourages the man to accept the need for change in the way he thinks and behaves in his marital relationship. This change requires the perpetrator to:

1. Face the abuse, including its extent and effects.

2. Recognize that the problem belongs to him rather than to others.

3. Respect the victim by creating appropriate boundaries between himself and the woman, including living separately temporarily.

4. Understand others' experience of his violence.

5. Understand how the pattern of violent behaviour he developed has prevented him from behaving more respectfully to others, insisting instead on satisfying his own needs.

6. Help the victim to avoid taking responsibility for his (the perpetrator's) actions.

7. Respect the rights of others, take on appropriate age, gender, and social roles and responsibilities.

8. Confront the consequences of his behaviour in terms of society, for example, legal proceedings.

9. Accept new domestic and social responsibilities and pressures without relapsing. The approach applied in a programatic way warrants evaluation.

Other cognitively-based models address the perpetrator's attitudes to women and his beliefs about the morality and usefulness of his violence (Ellis 1976); and assist him to articulate and change self-statements that occur automatically before, during, and after a violent episode (Deschner 1984). The models introduce more appropriate ways to deal with stress, anger, and aggression, and have been applied either individually or in a peer-group context.

An alternative point of view argues that too early an emphasis in treatment on violence *per se* leads to defensiveness in the patient who seeks to justify past actions (as well as those of his abusive parents if he was a victim of child abuse) (Koval *et al.* 1982). An impasse may ensue because the developmental egocentrism of such a patient impedes his moral reasoning. Accordingly this approach advocates a psycho-educational stance initially.

The clinician obtains a detailed account of relevant stressors and provides the patient with information about the physiology and psychology of stress. Techniques of self-care, self-control, and relaxation are taught. Sex-role stereotypes the patient holds about himself, his wife, and others are then addressed and their origins. The advantages and disadvantages of such views are discussed, especially the way he expresses his needs. The violent behaviour is examined in the context of all these factors, this setting the scene to practise more adaptive modes of expressing feelings, especially anger and vulnerability. Social and communication skills are also taught and practised.

Both approaches are cautious about focusing unduly the perpetrator's own history of childhood violence or abuse lest this encourages him to regard himself as a victim. Instead, treatment is self-focused, problem-oriented, with the perpetrator an active participant who observes, records, and experiments with his reactions and learns new skills in negotiation and disagreement.

Yet other approaches give more emphasis to psychodynamic factors (Walker 1983). For example, the man's sense of vulnerability, even if it is denied, renders him excessively dependent on the relationship with the woman, who then may be idealized. These dynamics, which have their parallels in social stereotypes of gender and marital relations, perpetuate the risk of violence. However it is clear that these issues should be addressed only after the the victim's safety is assured.

Conjoint or family therapy

When the risk of violence has diminished, when the perpetrator is attending his own therapy and showing a willingness and capacity to change, then therapy for both partners may be attempted, either with the couple alone or in a group context. Although research is limited experience suggests that the preceding individual therapy of each partner should continue.

A systems perspective notes the repetitive patterns of interaction which lead to escalating conflict. Individual processes underlying these patterns have been described in terms of psychoanalytic and behavioural theories and a clinical model integrating these approaches in family therapy has been proposed (Feldman 1982). The combination of structural family therapy with cognitive-behavioural approaches has also been applied (Taylor 1984).

CHILD ABUSE

Although both sexual and physical abuse receive considerable attention in both the professional and popular literature, many other forms of verbal and behavioural intimidation, humiliation, and physical and emotional neglect are inflicted on children. These types of abuse may coexist in one family and also be associated with other forms of family violence.

Because of limitations of space we focus on physical and sexual abuse only.

Physical abuse of children

Both men and women (usually a parent, caretaker, or cohabitee of a parent) commit violent acts against their children. When the time spent with children is allowed for it appears that men are more violent than women though the latter are over-represented in the research and clinical literature.

Kempe, an American paediatrician, introduced the term 'Battered Child Syndrome' in 1962 to describe the clinical presentation of young children who suffered severe, recurrent, and potentially fatal physical injuries at the hands of their parents or other adult caretakers. He estimated its incidence was 300 cases per million population per year, and argued that the appropriate medical, psychiatric, social, and legal care to protect the child at risk and to help re-establish a safe family environment was lacking.

Since Kempe, the helping professions and the public have become much more aware of this problem and of child sexual abuse. Although these two conditions, together with emotional neglect or emotional abuse of children are more likely to present to child psychiatrists, adult psychiatrists need to appreciate that some of their patients may be abusive parents. While certain psychiatric syndromes (for example, severe personality disorders and major mental illness) combined with adverse environmental circumstances are often associated with various forms of severe physical abuse, all adults are capable of causing physical injury to a child. This is particularly so in the context of the adult imposing discipline.

The perpetrator

Many of the individual, interpersonal, environmental, and family factors described in the section on marital violence also apply to the physical abuse of children. In addition the parents often are emotionally ill-prepared for, or impaired in their ability to function in a child-rearing role, sometimes to the extent of needing the child to validate the parents' otherwise-fragile self-esteem.

Diagnostic categories which increase the risk of a child abuse include: severe personality disorders, (including borderline, paranoid, anti-social, dependent, and immature); mood disorders, especially depressive illness; psychotic states, including post-partum psychosis, schizophrenia, and delusional disorder; alcohol and substance abuse and intellectual retardation.

Emotional unpreparedness for parenthood is particularly a feature of the adolescent parent, especially if she is a single mother who lacks emotional and financial supports. The recently divorced parent or parent in the throes of divorce especially if immature, and whose child is responding to the unhappy environment by behaving in a distressed or provocative manner, is potentially abusive. Often such parents have a very limited repertoire of disciplinary strategies which rapidly escalate into physical violence.

The victim

The child often has features that increase its risk of being physically abused or emotionally neglected. These include an unwanted or unexpected pregnancy, or one occurring in the context of interpersonal difficulties between the parents or with the extended family. As mentioned above, the child of a single parent is particularly at risk.

Congenital deformity, premature birth with attendant illness, or behavioural difficulties in early childhood are other factors, as is the socially unresponsive baby or young child or one with a low tolerance for frustration who cries a lot and sleeps poorly.

These features as well as the child's age-appropriate oppositional or experimental behaviour may disappoint the fantasies the parent needs to have about the child. The parent then feels a failure and becomes angry and violent towards the child who is perceived as wilfully defiant or prone to 'bad habits'.

Parent-child interaction

In the past decade considerable research into the neurophysiology and psychology of temperament has highlighted the relevance of an optimal 'goodness of fit' between the temperaments of child and parent, and the pattern of reciprocal cueing of emotional and behavioural responses between them. Discrepancies and disruption of these patterns occur more frequently in abusing than non-abusing families. Abusive parents tend to misread their child's cues and misinterpret his state of mind, often behaving in either an intrusive, controlling manner, or a detached fashion. Such parents have difficulty in differentiating between 'good' and 'bad' behaviour, or are over-inclusive in their definition of bad. Their threshold for frustration and expression of anger is relatively low. They also punish 'bad' behaviour to a greater degree than they reward 'good' behaviour by the child, and consider bad behaviour to be deliberate and wilful (Bolton and Bolton 1987)

The punitive parental attitude may lead to emotional detachment in the child, the two responses perpetuating one another. This has been implicated as a developmental factor in the patterns of social avoidance, hostility, and lack of sensitivity displayed by adults who were physically abused during childhood.

Family dynamics

The physically maltreating family has been described as a system dominated by power imbalances and by a tendency to react to feelings by immediate action (Bolton and Bolton 1987). The pattern of communication in such families is distorted, with much yelling and angry outbursts, but little resolution of most underlying conflicts. Members are either physically active and emotionally volatile towards one another or apathetic and oblivious of each other. A sense of ineffectiveness and unworthiness in the parent combined with emotional isolation within the marriage and from extended family and society may find expression in excessive anger toward the child, especially in the context of imposing discipline. A particularly salient family dynamic is that of the displacement of hostile feelings from the marital dyad onto the child.

The attempt to construct a typology of families in which violence toward children occurs has been prompted by the need to define those families in which it is relatively safe for the child as compared with those where only his removal can ensure safety. A review of such classification was inconclusive (Berger 1980).

Sexual abuse of children

Estimates of the prevalence of childhood sexual abuse range from 1 per cent – 15 per cent of the general population. In an adult psychiatric out-patient population up to 40 per cent report at least one episode of sexual abuse during childhood. One-third of reported cases are male. Most perpetrators are male.

Father-daughter incest is the most common pattern, 'father' including step-father, grandfather, uncle, mother's boyfriend, and other adult males considered members of the family circle. Adolescent perpetrators against younger siblings or siblings of friends is a category of particular concern which shares many features of father-daughter incest. Mother-son incest occurs much less frequently but its prevalence probably is under-reported.

As with physical abuse a wide range of behaviours is subsumed under the term sexual abuse, including intercourse, genital fondling, oral sex, voyeurism, and sexual innuendo. The degree of coercion and frequency of the abuse vary considerably. From a therapeutic point of view the victim's experience of the abuse is more important than its behavioural definition or frequency. This is particularly so if violence or the threat of violence accompanies the sexual abuse.

Features

Father-daughter incest usually begins between school age and puberty, although it may commence at a much younger age and continue to the teen years, even into early adulthood. 'Peeping' may occur initially and proceed gradually or quickly through stages of fondling, caressing, masturbation, oro-genital sex, and intercourse (vaginal or anal). Physical violence or threats against child or mother may occur, though sometimes the child intuitively senses the threat. Whether out of fear, misplaced trust, or lack of understanding of what is happening, the child rarely resists and submits to the adult's wishes. The likelihood of physical violence increases as the victim enters adolescent years and is then more apt to protest or threaten disclosure (Herman 1981).

Frequency and duration vary considerably, with over one-third of adult female victims reporting that their abusive experience regularly occurred for at least a year of their childhood. Among women suffering from alcoholism, substance abuse, or severe personality disorders the duration of their childhood experiences of sexual abuse commonly is reported as having lasted several years. Greater frequency and duration of abuse are associated with more severe psychological sequelae and similar severe consequences follow when the perpetrator was a close and trusted figure in the child's life.

Most information about the effects on children and adults is derived from studies of clinical populations. Factors that protect victims from developing difficulties of sufficient severity to warrant psychiatric treatment are not readily identifiable.

Clinical presentation

We now consider the range of presentations of sexual abuse in any patient.

In childhood, victims may present with sleep disturbance, nightmares, inappropriate anxiety and social avoidant behaviour, conduct disorders, anger and hostility, depression, and somatic complaints, especially abdominal and pelvic pain.

In adolescence a history of sexual abuse may underlie depression, attempted suicide, self-mutilation and other self-destructive behaviour, alcohol binges and drug abuse, eating disorders, running away from home, delinquency and persistent conflict with authority figures, chronic somatic complaints, conversion disorders, dissociative disorders, dysmorphophobia, psychotic states, precocious and promiscuous sexuality, venereal diseases, illegitimate pregnancy, and prostitution. The dissociative disorders are manifest along a spectrum of increasing severity ranging from an occasional dissociative episode, through post-traumatic stress disorder to the so-called multiple personality disorder.

In their adult years, in addition to the aforementioned problems in adolescence, victims of childhood sexual abuse may present with difficulties in forming stable, intimate relationships, sexual dysfunction, poor self-esteem, chronic depression, and self-destructive or self-defeating behaviours. In their relationships they continue to suffer and/or inflict physical, sexual, and emotional abuse on others. They may abuse or neglect their children, and women victims may be at increased risk of rape and coercive sex. This 'compulsion to repeat' the trauma of childhood has been studied extensively by appying the concepts of psychoanalysis, learning theory, ethology, and neurobiology (van der Kolk 1989).

Male victims may face particular problems regarding doubts about their masculine identity, homosexual fears, and a deep sense of shame and humiliation which prevent them from divulging the history of abuse. A proportion may have an increased risk of becoming child molesters (Groth 1978), as well as homosexual behaviour (Finkelhor 1984).

Many victims function reasonably well in their adult lives until they encounter a challenge which requires a change in the way they deal with intimacy, sexuality, or power relationships. Family life-events which may trigger such symptoms include the death of a parent or sibling, a child leaving home, a child becoming sexually active, or a child with whom the abused parent identifies strongly reaching the same age as the parent was when the abuse began.

Previously sexually abused women who suffer a sexual assault in adulthood (for example, stranger or marital rape) are at particular risk for long-term psychological impairment.

If the clinician is alert to possible incest in the patient's life and elicits a history by sensitive questioning, the patient's psychiatric state may actually worsen after the disclosure. Whenever she thinks about or discusses the abusive experience the patient may suffer much distress, vivid images, fear and dissociative states (the latter resembling the coping strategies she may have used during the incest experience). Nightmares, hallucinations, pseudo-hallucinations, recurrent intrusive memories, obsessional thoughts, and panic states may occur, as may feelings of great rage or helplessness. The sheer intensity of such feelings and the ways they are projected or retained as part of the sense of self may create particular problems for the patient and her therapist (Ellenson 1989).

Family features

In their extensive review of the subject Bolton and Bolton (1987) claim that the family in which child sexual abuse occurs often appears outwardly stable and adequately functioning. However the family is socially isolated, a state actively promoted by the male perpetrator, who may assert authority through temper outbursts, jealousy, and physical violence, more often directed against his wife than the daughter-victim. In structural terms there is a blurring of boundaries between individuals and between the sub-systems (especially father and daughter), disengagement between the parents, and in some cases between mother and daughter. The family's authority and executive functions are skewed and communication and role flexibility impaired.

As with other forms of family violence a description of the family in structural and systemic terms may be interpreted to mean that all members are equally responsible for the occurrence of the abuse. This view is unacceptable since it allows the perpetrator to avoid responsibility for his actions and to maintain a 'macho' sense of entitlement to use children (or women) to gratify his needs (Will 1989) (see Chapter 14). It also neglects the multiple levels of discourse (individual, family, societal) which construct gender relations (James and MacKinnon 1990).

Characteristics of the perpetrator

These are similar to perpetrators of other forms of abuse. The parental role brings many difficulties since he cannot discrimate easily between sexual and non-sexual affection. He seeks out a younger, smaller person to gratify his needs, in the pursuit of which he has been socialized to be active and persistent.

The nurture and care that he did not receive during childhood are now sought through relating to a child, and if his needs are not met or he is challenged he becomes enraged and may intimidate the family with violence or threats of abandonment. A wide range of psychological defence mechanisms also may be deployed, including massive denial, projection, and intellectualization.

Some researchers (Groth 1982) differentiate two groups of perpetrators. One group consists of paedophiles, whose psychosexual development is fixated on children, often compulsively. Although they may marry, this is in response to social pressures or to gain access to children (especially stepchildren). Perpetrators comprising the second group regress to a sexual relationship with children in the wake of conflict or disappointment in adult intimate relationships or other stress which undermines their sense of self.

The mother (non-abusive parent)

Society regards the mother of the sexually abused child as having failed to protect her child, and hence to have neglected the most fundamental task of her maternal role. Such 'mother blaming' is not justifiable since it implies that the husband is not entirely responsible for his actions. A typology of mothers who are not themselves perpetrators in sexually abusive families has been described (James and Nasjleti 1983):

1. Passive, childlike, immature, and dependent – she herself was a childhood victim of abuse, hates her own mother, and helplessly accepts that victimization (her own, her children's, and other people's) is an inescapable reality.

2. Emotionally detached, materialistically striving, and professionally successful – the husband is the nurturer and care provider in the family.

3. Rejecting and vindictive – she is hostile to and critical of her children,

refusing to acknowledge that her daughter may have been sexually abused, often blaming her instead.

Once again we emphasize that typologies and clinical descriptions are in no way intended to exonerate the perpetrator who has violated and betrayed his children. However, it is important to note that some victims, whether children or adults, describe their mothers as emotionally fragile, chronically burdened, emotionally absent, or in some other way unable to deal with the disclosure of the incest, and that this was a crucial factor in the reluctance to divulge it. On the other hand, it is also the case that many mothers, when informed of the abuse respond in an appropriate way to ensure their children's safety (Johnson 1992).

Characteristics of the victim

As described above, in some families the abuse, especially of daughters, occurs serially. In other families only one child is 'selected', but it is unclear what features in the child, if any, facilitate this. Heightened sensitivity to the feelings and needs of others, especially those of the perpetrator, is widely reported. So too are the closely related phenomena of 'pseudo-maturity', 'knowingness', and 'feeling different from other children'. However, the question remains whether these characteristics are the cause or the result of the child's premature and inappropriate introduction to the world of adult sexuality. Furthermore, the child's role commonly has included a sense of responsibility for the welfare of the family in general, to which is then added a concern about the consequences of disclosure for family, perpetrator, and the child herself. This quandary increases her sense of isolation and vulnerability to further exploitation. Even when the secret is shared with a sibling (who may also be a victim), shame combined with fear of the father's tyranny and the effects of disclosure on the family aggravate the victim's isolation and her reluctance to seek help.

Treatment of physical and sexual abuse of children

Principles of treatment include: (a) ensuring the safety of the child and other potential victims; (b) treatment for the offender; (c) treatment for the child victim(s); (d) therapy for the marital couple, and (e) family therapy.

As well as individual, couple, and family therapy the treatment approaches may be applied in a group context.

The clinician who is involved in treatment necessarily contends with several emotionally charged factors which may complicate an already problematic treatment. These include:

1. The impact of disclosure on the family. Particularly in cases of incest the disclosure may lead to profound reactions in members toward one another, the victim and the helping professionals. Reactions include:

(a) The parents perceive the disclosure as a threat to their marriage and 'close ranks', the clinician then entering into an escalating struggle with them to validate the victim's point of view. The greater the parental reluctance, the more the clinician becomes an advocate for the victim and is drawn into blaming the parents.

(b) Alternatively, mother and victim may appear to join forces and link up with the clinician. All agree that the father is to be banished, with consequent guilt by the daughter for disclosing the incest and thus destroying the family. The clinician may be tempted to identify with the expelled father, and feels that daughter or mother have acted vengefully. Antagonism may then develop between the clinician and mother, or between clinician and victim, the latter again experiencing a sense of betrayal.

(c) The victim may retract her allegation, an act supported by both her parents.

If treatment is conducted by a family therapy team, its members may be divided over the issue of innocence or guilt. A split is particularly likely between the clinician treating the patient and the other professionals who may adopt a more systemic perspective. Such conflicts and divisions among the helpers commonly mirror the interpersonal and intra psychic dynamics of the family members.

2. A second complicating factor is *the clinician's possible conflicting roles* as therapist and agent of social surveillance. Even if he is able to hand over responsibility for the victim's safety to relevant social agencies, she and other family members may remain suspicious of the clinician who they believe has the power to break up the family and send the father to jail. Likewise the social agency which is mistrustful of the parents grows increasingly authoritarian towards them and the therapist. One approach to the difficult task of responding to these competing demands has been provided by Mackinnon and James (1991, 1992*a*, 1992*b*) (see below).

3. Following disclosure the medical and social *investigatory process* and the judicial inquiry that sometimes follows may be traumatic for the victim.

Principles of treatment
Protection of the victim and potential victims from further abuse is the first essential step. As discussed earlier the clinician's role must be differentiated from that of the victim's guardian. Mackinnon and James (1991, 1992*a*, 1992*b*) emphasize the importance of winning the family's trust and that of statutory welfare services before formal family therapy can begin. The

concerns of the protagonists are empathically acknowledged, circular questioning and written contracts may be used to identify the goals of treatment and criteria of change desired by the various parties, and the clinician actively and openly negotiates with all groups to effect these changes.

The principles of *treatment of the perpetrator* of physical or sexual abuse resemble those described in the corresponding section on marital violence. In addition, a psycho-educational approach for both adult partners on such matters as the stages of child development and appropriate ways of handling these, provides useful information and skills which reduce the likelihood of further abuse. Alongside this, a range of psychotherapeutic methods are available to help the perpetrator develop a more robust sense of self, one which is not reliant upon the child. The perpetrator is required to accept responsibility for his actions (Jenkins 1990, 1991). Indeed, some clinicians advocate that he must formally acknowledge his violation of his child's trust (Trepper 1986) (see Chapter 14). The emphasis then shifts to his gaining control over his behaviour.

Individual *therapy for the non-abusive spouse* may be indicated, particularly in a case of child sexual abuse, where she may need to explore her own feelings about the incest in the light of her own history and of the beliefs she has had about her marriage and role as parent. Her conflict of loyalties towards victim and perpetrator often defies a straightforward solution, and the clinician should be wary of attempting to provide one. A sensitive description of how therapy may help resolve this dilemma is given by Sheinberg (1992).

In addition to individual therapy, the spouse may have joint sessions with the victim, and, separately, joint sessions with the perpetrator. She also may need support in coping with the distress of marital separation and the burdens of single parenthood if the safety of the children cannot be achieved by other means.

Siblings of the victim should be assessed individually, not only for evidence of abuse but for their own reactions to the disclosure and concerns for other family members.

The specific individual treatment of the child victim of sexual abuse is beyond the scope of this book; many useful reviews are available (see, for example, Sgroi 1982; Kempe and Kempe 1984).

As with most forms of therapy which are used in the management of family violence, *family therapy* requires modifications in the context of the physical and sexual abuse of children. These include integration of community support and legal agencies with a multifaceted treatment programme (Giarretto 1982), and a step-wise goal-oriented treatment programme for all family members (James and Nasjleti 1983).

These programmes, while urgently needed and often enthusiastically endorsed, have not been adequately evaluated. The paedophile-type of abuse in particular may only be solved by the perpetrator's separation from the family and possible incarceration.

Treatment of *adults who were childhood victims* of sexual abuse have been described from a number of perspectives, including individual psychodynamic (Stone 1989), individual, systems-oriented, feminist (Sanders 1992), family systems-oriented feminist (Barrett *et al.* 1990), cybernetic-contextual approaches for individual and families (Durrant and White 1990), and group therapy (Goodwin and Talwar 1989).

The effectiveness of these approaches for such a difficult and heterogeneous group of patients awaits evaluation.

ELDER ABUSE

The incidence of elder abuse in the USA is estimated at 10 per cent of the population over the age of 65 years, almost half of the cases classified as moderately severe or greater in their degree of seriousness (American Medical Association 1990).

A broad range of abuse occurs in the home or institutions, including assault; forcible restraint; inadequate nutrition and clothing; neglect of the need for spectacles, hearing aids, false teeth, and walking frames; lack of privacy; and social isolation. Inadequate, inappropriate, or unsupervised medication is a particularly worrying problem.

This subject is discussed more fully in Chapter 12, p.239.

REFERENCES

American Medical Association (1990). American Medical Association White Paper on Elderly Health. Report of the Council on Scientific Matters. *Archives of Internal Medicine*, **150**, 2459–72.

Barrett, M.J., Trepper, T.S., and Fish, L.S. (1990). Feminist-informed family therapy for the treatment of intra family sexual abuse. *Journal of Family Psychology*, **4**, 151–66.

Berger, A.M. (1980). The child abusing family: II child and child-rearing variables, environmental factors and typologies of abusing families. *American Journal of Family Therapy*, **8**, 52–68.

Black, D. and Kaplan, T. (1988). Father kills mother: issues and problems encountered by a child psychiatrist service. *British Journal of Psychiatry*, **153**, 624–30.

Bograd, M. (1984). Family systems approaches to wife battering: a feminist critique. *American Journal of Orthopsychiatry*, **54**, 558–68.

Bograd, M. (1992). Values in conflict: challenge to family therapists' thinking. *Journal of Marital and Family Therapy*, **18**, 221–38.

Bolton, F. and Bolton, S.R. (1987). *Working with violent families*. Sage, Newbury Park, CA and London.

Carlile, J.B. (1991). Spouse assault on mentally disordered wives. *Canadian Journal of Psychiatry*, **36**, 265–9.

Cate, R.M., Henton, J.M., Koval, J. *et al.* (1982). Premarital abuse. A sociological perspective. *Journal of Family Issues*, **3**, 79–90.

Deschner, J.P. (1984). *The hitting habit: anger control for battering couples*. Free Press, New York.

Dickstein, L.J. (1988). Spouse abuse and other domestic violence. *Psychiatric Clinics of North America*, **11**, 611–28.

Durrant, M. and White, C. (ed.) (1990). *Ideas for therapy with sexual abuse*. Dulwich Centre Publications, Adelaide.

Ellenson, G.S. (1989). Horror, rage and defenses in the symptoms of female sexual abuse survivors. *Social Casework*, **70**, 589–96.

Ellis, A. (1976). Techniques for handling anger in marriage. *Journal of Marriage and Family Counselling*, **2**, 305–15.

Feldman, L.B. (1982). Dysfunctional marital conflict: an integrative interpersonal-intrapsychic model. *Journal of Marital and Family Therapy*, **44**, 417–28.

Finkelhor, D. (1984). *Child sexual abuse: new theory and research*. Free Press, New York.

Frude, N. (1991). *Understanding family problems – a psychological approach*. Wiley, New York.

Giarretto, H. (1982). *Integrated treatment of child sexual abuse: a treatment and training manual*. Science and Behaviour Books, Palo Alto, C.A.

Gillman, I.S. (1980). An object-relations approach to the phenomenon and treatment of battered women. *Psychiatry*, **43**, 346–58.

Goldner, V., Penn, P., Sheinberg, M., and Walker, G. (1990). Love and violence: gender paradoxes in volatile attachments. *Family Process*, **29**, 343–64.

Goodwin, J.M. and Talwar, N. (1989). Group therapy for victims of incest. *Psychiatric Clinics of North America*, **12**, 279–93.

Groth, A.N. (1978). Sexual trauma in the life histories of rapists and child molesters. *Victimology*, **4**, 10–16.

Groth, A.N. (1982). The incest offender. In *Handbook of clinical intervention in child sexual abuse* (ed. S.M. Sgroi) Lexington Books, Lexington MA.

Herman, J. (1981). *Father-Daughter incest*. Harvard University Press, Cambridge, MA.

Hilberman, E. (1980). Overview: the 'wife-beater's wife' reconsidered. *American Journal of Psychiatry*, **137**, 1336–47.

Hughes, H.H. (1988). Psychological and behavioural correlates of family violence in child witnesses and victims. *American Journal of Orthopsychiatry*, **58**, 77–90.

James, B. and Nasjleti, M. (1983). *Treating sexually abused children and their families*. Consulting Psychologist Press, Palo Alto, CA..

James, K. and Mackinnon, L. (1990). The 'incestuous family' revisited: a critical analysis of family therapy myths. *Journal of Marital and Family Therapy*, **16**, 71–88.

Jenkins, A. (1990). *Invitations to responsibility*. Dulwich Centre Publications, Adelaide.

Jenkins, A. (1991). Intervention with violence and abuse in families: the inadvertent perpetuation of irresponsible behaviour. *Australian and New Journal of Family Therapy*, **12**, 186–95.

Johnson, J.J. (1992). *Mothers of incest survivors – another side of the story*. Indiana University Press, Bloomington and Indianapolis.

Kempe, R.S. and Kempe, C.H. (1984). *The common secret: sexual abuse of children and adolescents*. W.H. Freeman, New York.

van der Kolk, B. (1989). Compulsion to repeat the trauma: re-enactment, revictimisation and masochism. *Psychiatric Clinics of North America*, **12**, 389–411.

Koval, J.E., Ponzetti, J.J., and Cate, R.M. (1982). Programmatic intervention for men involved in conjugal violence. *Family Therapy*, **9**, 147–54.

Mackinnon, L. and James, K. (1991). Initial meetings in child-at-risk cases: developing a therapist–family alliance. *Australian and New Zealand Journal of Family Therapy*, **12**, 175–85.

Mackinnon, L. and James, K. (1992*a*). Working with 'the welfare' in child-at-risk cases. *Australian and New Zealand Journal of Family Therapy*, **13**, 1–15.

Mackinnon, L. and James, K. (1992*b*). Raising the stakes in child-at-risk cases: eliciting and maintaining patients' motivation. *Australian and New Zealand Journal of Family Therapy*, **13**, 59–71.

Pagelow, M.D. (1984). *Family violence*. Praeger, New York.

Russell, D.E.H. (1982). *Rape in marriage*. MacMillan, New York.

Sanders, C. (1992). 'A long road home' – working with adult survivors of child abuse from a systemic perspective. *Australian and New Zealand Journal of Family Therapy*, **13**, 16–25.

Sgroi, S. (1982). Handbook of clinical intervention in child sexual abuse. Lexington Books, Lexington, MA.

Sherman, L.W. and Berk, R.A. (1984). The Minneapolis domestic violence experiment. *Police Foundation Report*, **1**, 1–8. Quoted in Bolton, F. and Bolton, S.R. (1987). *Working with violent families*. Sage, Newbury Park CA and London.

Sheinberg, M. (1992). Navigating treatment impasses at the disclosure of incest: combining ideas from feminism and social constructionism. *Family Process*, **31**, 201–16.

Stone, M.H. (1989). Individual psychotherapy with victims of incest. *Psychiatric Clinics of North America*, **12**, 237–55.

Stulberg, F.L. (1989). Spouse abuse: an ecosystemic approach. *Contemporary Family Therapy*, **11**, 45–60.

Taylor, J.W. (1984). Structured conjoint therapy for spouse abuse cases. *Social Casework*, **65**, 11–18.

Trepper, T.S. (1986). The apology session. In *Treating incest: a multiple systems perspective* (ed. T.S. Trepper and M.J. Barrett). Haworth Press, New York.

Walker, L.E. (1979). *The battered woman*. Harper Colophon, New York.

Walker, L.E. (1983). The battered woman syndrome study. In *The dark side of families* (ed. D. Finkelhor, R. Gelles, G.T. Hotaling, and M.A. Straus) Sage, Beverley Hills CA.

Weingourt, R. (1985). Never to be alone. Existential therapy with battered women. *Journal of Psychosocial Nursing*, **23**, 24–9.

Will, D. (1989). Feminism, child sexual abuse and the (long overdue) demise of systems mysticism. *Context*, **1**, 12–15.

Yllo, K. and Strauss, M.A. (1981). Interpersonal violence among married and cohabiting couples. *Family Relations*, **30**, 339–47.

14

Ethics and the family

This final chapter on ethics and the family is by no means a tag. We hope that the relevance of the topic may have become self-evident from our many allusions to it in preceding chapters. We would contend that multiple issues of an ethical nature are pertinent when: thinking about the family as a system; determining when the system becomes dysfunctional; assessing the clinical problems of the presenting patient in the family context; and trying to devise an optimal, therapeutic program for a family. Several ethical topics warrant our attention at each level; we have opted to concentrate on those which are particularly relevant to themes covered in this book.

We will not dwell therefore on aspects, albeit important, which are more appropriately placed in a text on family therapy *per se* (for example, the use of paradoxical intervention, which some critics regard as deception; training and competence; and the role of a code of ethics for family workers). We focus instead on matters central to our remit: the relevance of ethics in constructing models of family function and dysfunction; the salience of values in family assessment including the complex question of how to satisfy the interests of all family members; confidentiality; and informed consent.

ETHICS AS THE 'CORNERSTONE'

We trust the following discussion of the role of values in the clinical approach to the family will persuade the reader that to ignore this dimension is tantamount to committing a disservice to the family. Some theorists grasp the nettle firmly by: (a) making this ethical dimension explicit; and (b) elevating it to a central position in their model of the family. The most notable example is Ivan Boszormenyi-Nagy (Boszormenyi-Nagy *et al.* 1991) and it is to his ideas that we turn in order to illustrate key features in the interface between ethics and professional intervention with the family.

One of the four dimensions of his model (the other three are not pertinent to our purpose), entitled 'relational ethics: the balance of fairness', is regarded as a 'cornerstone' since it is based on the uniquely human process of a group of people seeking an 'equitable balance of fairness' among themselves; the process is one based on trust, 'an essential fact of human existence'. Within a family group too, trust together with its associate, trustworthiness, facilitates an exchange of mutual obligations in which the interests of each member are acknowledged and respected by the rest of the family. The picture is complex in that members have both debts and legitimate entitlements, the latter gained by contributing to the well-being of others.

The resultant family 'ledger' is an account of what has been given by and what is owed to various members. The ledger is markedly affected by previous generational factors and the stage of the family life cycle. For instance, in the 'young child' nuclear family, an equitable asymmetry prevails, with parents the most responsible for giving, mainly to their dependent offspring. At the parental level itself, the marital relationship embodies a symmetry of rights and responsibilities but also asymmetries (for example, acknowledged male-female differences concerning repro-duction and nurturance) which are dealt with by agreed-upon assignment of roles to meet the family's needs. Children mainly respond to parents' giving through a sense of loyalty to their family.

The family optimally accomplishes a balance of fairness through a just distribution of burdens and benefits, including the redressing of previous injustice. Rather than permit such injustice to be ignored, thus leading to ethically invalid 'solutions' and corresponding dysfunction (for example, a sense of 'destructive entitlement' in a family member causes him to relate maliciously towards innocent others), the family harnesses its resources of trustworthiness with stress on mutual care and concern. Despite any 'predicaments' the children may have inherited, and given that life cannot be lived over again, they may need to exonerate or forgive previous generations; this contribution is part of their constructive, ethical potential.

Boszormenyi-Nagy does not confine himself to issues of fairness intrinsic to the family. Broader, social forces like ethnicity, socio-economic status, and gender are linked to special forms of fairness. For example, cultural norms may be more or less just – a contemporary illustration often raised by feminist theorists is the detrimental effects of sexism and roles imposed on the wife in the conventional family (see Chapters 1 and 12).

This brief account should suffice to highlight the pre-eminent place for ethics in a professional approach to the family. Major repercussions follow for clinicians which extend well beyond the customary caveats that they should be aware of the pervasiveness of values in the clinical encounter as well as the risk of imposing their own values (Bloch 1989; Holmes and Lindley 1989). According to Boszormenyi-Nagy, we should participate actively and explicitly in reframing the patient's presenting problems into ethically bound themes, these then supplanting conventional, clinical constructs.

Some representatives of this position insist upon even more robust, ethical intervention. Let us consider racism as an illustration. Charles Waldegrave and his group (1991–92) (including Maori, Samoan, and White professionals) in the Wellington Family Centre in New Zealand have accorded paramount importance to race in their family therapy and related community development work (also to gender and poverty). They operate on the premise that the problems of many families are 'imposed by broader social structures' rather than being inherently familial. Foremost among them is the long-standing unfairness suffered by the indigenous Maori population, a group historically marginalized by the dominant White culture. The family model consequent upon such reflections is dubbed 'Just therapy' in order to convey a sense of its principal aim – to highlight the injustice meted out to the Maori family and the need for it to be redressed. This could be misinterpreted as paternalism but this is not the case for the Wellington group. Instead, the clinical encounter 'relates to the manner in which people give meaning to experience and create their 'reality'. Thus, the family itself examines its current reality which necessarily embodies a social and political agenda as well as an intrafamilial one.

A similar view is held by two South African psychologists working with Black families in the Child Guidance Clinic of the University of Cape Town. Steere and Dowdall (1990) have argued that assessment and treatment of these families' problems requires both a deep awareness and disclosure by the clinician of his own political position. Given the blatant, societal racism in South Africa, effective intervention is contingent upon an appreciation of the deleterious sequelae of apartheid on family life and a corresponding commitment to declare one's fervent denunciation of it (in 1985, for example, the South African Institute for Clinical Psychology publicly condemned apartheid as unethical). Thus an open avowal of one's political stance on apartheid becomes a prerequisite of ethically guided intervention.

A MATTER FOR DEBATE

Whilst the positions taken by Boszormenyi-Nagy, Waldegrave *et al.*, and Steere and Dowdall appear coherent and legitimate, they are subject to debate. Boszormenyi-Nagy's views for instance have been attacked by Wendorf and Wendorf (1985), their criticism partly theoretical, partly pragmatic. His notions, they assert, are contrary to systems theory because of their linearity, unidirectionality, and individualism; specifically, the family is construed as a collection of hostile, guilt-ridden, isolated persons enveloped solely in conflict. Boszormenyi-Nagy allegedly ignores the mutuality, reciprocity, and complementarity of family relationships, and the associated advantages of family life. Practically, this 'individual ethics' approach places undue limitations on family assessment and treatment; and also unjustifiably calls on the clinician to advance a political cause.

A close study of the Wendorfs' argument reveals its rather rickety basis and its lack of soundly conceived premises. Nonetheless, it remains indisputable that Boszormenyi-Nagy's elevation of 'relational ethics' to paramount status signifies a radical shift in construing, and intervening with, the family.

THE SEXUAL ABUSE CONTROVERSY

At another level of the 'family-ethics' debate, protagonists focus on circumscribed themes although the issues are generalizable. Consider the example of child sexual abuse (CSA), an exceedingly critical matter in itself but one which lends itself well to the appraisal of the place of ethics in dealing with the family. An illuminating dialogue between two British psychiatrists, David Will (1989) and Arnon Bentovim (1989) raises the relevant issues splendidly.

Will launched the exchange with an attack on the conceptual understanding of CSA by Bentovim and his team (then working in the pioneering CSA clinic of the Great Ormond Street Hospital in London). The gist of their position, usefully summarized by Bentovim *et al.* (1988) is that CSA is a family affair, best viewed within the framework of General Systems Theory, particularly circular causality (see Chapter 3). Thus, a father (or stepfather or other father substitute) initiates a sexual tie with a child; the degree of intimacy grows into gentle touching; the child, sexualized by the contact, responds sexually; the father then rationalizes this response as a justification to continue the intimacy; an interactive pattern becomes established based on circular causality; associated marital (and usually sexual) difficulty contributes to the cycle with continuing reinforcement thereafter.

Lustig *et al.* (1966) originally ushered in this type of thinking, regarding family incest as a means of maintaining family equilibrium when the sexual relationship between the parents had collapsed. One or more daughters were recruited to replace the sexually reluctant but colluding wife as the husband's partner/s. The results were a wife able to withdraw from her husband's sexual demands and the preservation of the family system.

What did Will oppose in Bentovim's model? For several years, it seems, nothing at all. Indeed, he had applied the approach in his own work with CSA families. But his continuing experience buttressed by the findings of empirical research eventually persuaded him that: '. . . the family system's view of CSA [was] profoundly wrong and ultimately dangerous'. Research data, he asserted, had (a) failed to support the notion that mothers collude with incest; and (b) revealed minimal differences between intra- and extra-familial sexual offenders. Moreover, the work at Great Ormond Street was uncontrolled, and limited by a substantial drop-out rate. Given this knowledge, it was impermissible to continue to promote a systems model. Instead, recognition should be paid to an amalgam of factors including psychoanalytic, personality traits in the perpetrator, and sociocultural. As a corollary, clinicians were obliged to accept the centrality of the perpetrator's role in CSA. Bentovim (1989) predictably issued a rebuttal to Will's attack; both positions were legitimate; the systems model had something valuable to contribute to the understanding of CSA and this did not negate other views focused on the perpetrator. Although not a fudge, the retort still did leave the debate hanging. The pivotal, ethical question whether men who sexually abuse their children are morally culpable also remained unanswered.

Other contributors to the CSA field have taken the 'ethical plunge' by introducing an unambiguous moral dimension into their work. Chloe Madanes (Carr 1989), a prominent representative of the strategic therapy school, has gone so far as to invoke the need for repentance and cor-responding forgiveness. Thus, for instance, the perpetrator is indubitably branded as offender, the abused as victim. Notwithstanding that both (and, indeed, other family members) have suffered spiritual pain as a consequence of the experience, the offender remains culpable and it is he who, morally bound, must repent in the presence of the victim, make a long-term, symbolic sacrifice, and learn how to forgive himself.

These requirements as well as the ritual of repentance embody a quasi-religious element, certainly an embrace of the ethical.

The school of 'abuser as culprit' obviously may entail a position in which blurring of the central issue is obviated by a distinction made between abuser and abused, with specific management of each. This clarification of roles also permits a clear perception of the relative power accompanying such roles – the adult male occupies an entrenched, patriarchal power base whereas the child is not only powerless but in no position to provide or withhold consent. The wife, like the abused child, cannot be held to account for her husband's behaviour no matter the atti-tudes she bears towards him, their sexual relationship, or the family as a whole.

A considerable literature has grown in recent years covering the ethical contours of CSA (and other forms of family violence such as physical abuse and wife battering (see Chapter 12); the interested reader will find the contributions by Goldner (1985, 1988), Leupnitz (1988), and Doherty and Boss (1991) useful.

'NEGOTIATING VALUES' WHEN WORKING WITH THE FAMILY

This subtitle is derived from an influential paper by Aponte (1985) entitled 'The negotiation of values in therapy'. Although his focus is on therapy *per se*, many of his points relate closely to a theme pervading this chapter, namely the central place of ethics in the way the family is construed. This is especially the case in terms of the polarity of well-functioning versus dysfunctional and its corollary of how we should assess the clinical problems of a presenting patient when family factors appear relevant. Such factors pertain on most occasions according to the way we argue in this book.

Aponte (1985) sums up his thesis (with which we entirely agree but modify to suit our own purposes) this way:

Values are integral to all social systematic operations and therefore to the heart of the therapeutic process [and to the conceptual and assessment process in terms of our present discussion]. For the therapist values are an essential component in defining and assessing a problem, determining goals, and selecting therapeutic strategy. Therapists do not have a choice about whether they need to deal with values in therapy [and also in assessment], only how well.

Elsewhere Aponte avers: 'The task . . . is how therapists can work with their professional and personal values in ways that benefit the families they treat'. (See Note 1 at the end of the chapter).

VALUES AND THE FAMILY IN CLINICAL PSYCHIATRY

Readers who have tackled Note 1 on the philosophical discussion about values may well be inclined to throw in the towel by now, hankering for a return to more familiar territory. Let us therefore forge a link between values (whatever the definition and corresponding theory preferred) and the clinical encounter with the family. An initial step is to recognize their interconnectedness. Secondly, some values are customarily espoused by the family as a group whereas others are held by members individually;

of course many variations occur from individual members' staunchly avowed separate positions to an undifferentiated adherence to a shared set of values.

Thirdly, the values concerned are far from being mundane or banal; indeed, they may be of the utmost significance for the family, representing that which ultimately binds or imperils them as a social system.

Fourthly, the values commonly interdigitate with the problems of the presenting patient and/or the associated systemic problems intrinsic to the family (for a clear-cut example we need only but return to the aforementioned debate on CSA).

Fifthly, the professional joining the 'clinical family' – no matter how beneficent his strivings or determined his inclination toward neutrality – will be reminded by the members of his own value judgements in relation to the clinical agenda. Finally, he will have to pay special heed to those values, checking their content and potential effect on his involvement as a clinician (and as therapist if treatment is embarked upon).

This last requirement is akin to a therapist's ongoing sensitivity to countertransference and its potential to enhance or hinder the clinical endeavour (this takes us into the realm of treatment and beyond the scope of the book; for a good account of the subject see Lakin (1988), Walrond-Skinner and Watson (1987), and Holmes and Lindley (1989)).

We are now equipped to return to Aponte's paper and to tackle the question of how values are optimally dealt with in approaching the family. We interweave our own views given their close resemblance to those of Aponte. His stress on the professional's 'primary responsibility to assist the family with the improvement of function' is a salutary reminder of the closely interrelated *desiderata* of professional intervention:

(1) we do not usurp the family's role as agent in grappling with its own problems both in terms of construing those problems and then attempting to remedy them;

(2) we take care to collaborate in this process, sharing our expertise when this is apt, but otherwise mainly encouraging the family to exploit its own assets and resources;

(3) we refrain from imposing our values on the family; thus, we resist declaring what is 'good' or 'right' or 'desirable' for them;

(4) on the other hand, we *do* point out, if necessary, the need for reappraisal of its values in order that corresponding structural and functional change can be achieved (for example, that a husband's

physical abuse of his children is a severe impediment to the family's potential for 'balanced fairness' (to cite Boszormenyi-Nagy));

(5) we discreetly point to disparities in the espousal of values *between* family members in the light of a judgement that this is a likely source of conflict (for example, between parents and children in relation to 'social drug experimentation') as well as the foundation of other dysfunction.

These steps are more easily stipulated than implemented. We are, after all, heavily invested in our value system, customarily long in the making and often derived from profoundly influential sources (like parents, grand-parents, and cherished teachers). Moreover, we may stumble in trying to keep them out of the consulting room in the face of a family's set of values which appears neither in its interests nor conducive to a necessary reappraisal preceding commitment to change. A keen, personal awareness of our vulnerability in this regard is pivotal, aided by feedback from peers if a professional colleagueship is available; both depend on honest self-scrutiny.

The above caveat and its proposed management are nicely illustrated by Asen and his colleagues (1991) when wrestling with their disagreement over appropriate criteria of improvement in families participating in a treatment trial. Since their exchange in terms of values parallels the issues we have raised and their attitude is one of refreshing candour, let us consider the entanglements by which they were discomforted.

In a study of 18 families with a wide range of presenting problems, there were substantial differences among the researchers as to the inter-pretation of the outcome data. The decision had been taken to employ a multidimensional approach, facilitating the assessment of change at the individual, dyadic, and family system level. At follow-up, improvement was noted on the individual and dyadic scales but not on the family system measure. This last scale included ratings of such family dimensions as communication, boundaries, alliances, adaptability, and competence.

Given the outcome, it is not surprising that the researchers interpreted the data in diverse ways, for example. (a) there was no change in overall family functioning; (b) the family system measure was a trait measure and therefore non-reactive to treatment; and (c) the model of family therapy used in the first place was inappropriate.

One could not ask for a greater diversity of views. Asen *et al.* were particularly insightful in concluding that whereas their study had set out to test the utility of different outcome measures, 'in the event, what was on trial were the assumptive worlds of the clinicians and of the researchers'.

The experience of the group is illuminating for all mental health professionals working with families since it raises the profound question of what constitutes proper goals of intervention. Also emerging from their exchange is the related issue of how we balance the needs of the presenting patient and those of other family members; it is this important topic that we now discuss.

THE RELATIVE NEEDS OF FAMILY MEMBERS INCLUDING THE PRESENTING PATIENT

Competing premises are relied upon in tackling the question of how we deal with the relative needs of the presenting patient and other family members. They include the following:

1. The needs and interests of the patient are paramount and constitute the priority target in the professional's endeavour; the family's involvement is ancillary, albeit important. Its resources are garnered but only in relation to the interests of the patient.

2. The needs and interests of family members are *equally pertinent* in that the family as a system is inescapably bound up with the patient (see Bloch *et al*. (1991) for illustrations of this position). All members are rigorously attended to in the clinical encounter although their level of involvement and suffering may vary from minimal to profound. Clinically and ethically, each member is a focus for the professional who resists alliances with an individual or sub-group (except transiently as a strategy to win trust; see Palazzoli *et al*. 1980).

3. The needs and interests of all family members are pertinent but *unequally* so, depending on particular circumstances. Pragmatism prevails but not of an arbitrary or idiosyncratic type. Members occupy different roles such as the vulnerable, the source of support, the potential beneficiary, the quasi-therapist therapist, and the caregiver as sanctioned by the family.

The presenting patient is not necessarily the clinician's chief priority. Indeed, he may be functioning relatively well but at the expense of a fellow member. To illustrate, a designated caregiver whose own needs are unintentionally neglected by the system may be floundering and consequently in need of aid herself. Moreover, no member necessarily occupies the same role continuously; needs may escalate or diminish with changing clinical and family circumstances.

Advantages and disadvantages, both clinical and ethical, typify each of these three positions. We would be foolhardy to argue the merit of one to the exclusion of the others. The preceding chapters, we hope, would have confirmed the potential for each position dependent on the special situation of the patient and the family. But, we hasten to add – we do not opt for nebulous eclecticism, for judgements based on whim or half-baked hypothesis. Indeed, we spell out our position in the following schematic way.

In promoting the virtue of the biopsychosocial model to patients and their families, with its inherent emphasis on the relevance of family factors in aetiology, prognosis, and treatment, we not only offer a clinical model but also convey our commitment to an ethical position. Let us elaborate. We indicate implicitly or directly that psychiatrists must necessarily engage the family in order to discharge their professional responsibilities. In manifesting this commitment, we call for a corresponding response from the family based on their trust in our expertise. But a corollary follows. We have an obligation to the family to ensure that their concerns and interests (but not necessarily their wishes, which may be unrealistic and discordant with reasonable clinical goals) are dealt with satisfactorily. In practical, ethical terms, we owe this to every member recruited to the clinical task, whatever their status (for example, presenting patient or not, adult or child, psychologically vulnerable or robust) and whatever values they espouse. At the least we strenuously avoid harm befalling them (according to the principle of non-maleficence or *primum non nocere*); optimally, we endeavour to see that their best interests are served.

So far, so good. Despite our best intentions however, the encounter may be fraught with difficulty as the family or one or more of its members resist our efforts, whether on rational or irrational grounds, and place our initiative in jeopardy, even to the extent of destroying it. The most crucial strategy to pre-empt this is to elevate the process of informed consent (highlighted below) into a central feature of our 'negotiations' (to use Aponte's term again) and to keep this in the forefront throughout our clinical involvement. Obviously, consent may be refused initially or withdrawn later but then this possibility applies to all psychiatric work.

The novel, potential complication with family recruitment is the need to obtain consent from several persons including perhaps minors.

No problems occur when the family is in accord with what is proposed and then enacted. The members and professional are as one, freely sharing common objectives. The situation however is inevitably more complex when members disagree with one another about whether or not to participate. Here, the call is for a reconciliation of views but not at the expense of anyone being badgered or harassed to comply; failing that, an invitation is extended to the family to explore the divisiveness in an effort to understand and possibly surmount it.

In both cases, the professional's self-scrutiny, especially of his values, is mandatory lest neutrality is jettisoned with biased judgement ensuing.

Of the three aforementioned positions we advocate the last, that is a respect for all family members' interests without any person being accorded priority merely by dint of his status at the time of the family assessment (status as patient or not is the most troublesome polarity). In essence, we argue that respect for the autonomy of every member is maintained, each having an absolute moral value, being regarded as an end in himself, and never seen merely as an means to an end (Kant's (1973) formulations are pivotal in this context).

Progression from family recruitment and initial negotiation of purpose and strategy *vis-à-vis* assessment to a therapeutic encounter brings with it a host of additional value-laden issues. These would take us beyond our remit; the interested reader is referred to the reference list at the end of the chapter.

Finally, we focus on confidentiality and informed consent, two matters of considerable ethical import for the clinician working with families.

CONFIDENTIALITY

If the issue of preserving confidences in the assessment of the individual for psychotherapy is problematic, then the matter is considerably more involved and, finally equivocal, in the case of a family (see Doherty and Boss 1991; Lakin 1988; Fieldsteel 1982; and Goldenberg and Goldenberg 1991). The typical assessment procedure involving a patient and his family invariably entails the disclosure of highly personal information. On occasion, such information is readily shared among all members since the discloser (whether patient or other family member) believes this is in the best interests of the patient and/or the family. By contrast, other information may be kept tightly concealed lest its disclosure should cause offence or other harm.

Not uncommonly, the clinician is confronted by the latter dilemma; in the course of interviewing the patient, or a family member in the role of informant, they may request or indeed insist that certain matters should under no circumstances be divulged to any party including the family. These confidences may suggest themselves as crucial to any subsequent family assessment (examples include sensitive topics like a history of sexual or physical abuse, sibling incest, wife battering, and an extra-marital affair). The clinician is then awkwardly placed. On the one hand, the shared material is in all likelihood at the heart of the family dynamics and presumably therefore in need of open, collaborative consideration and analysis. On the other hand, the potential to achieve a trusting, working alliance with the patient is jeopardized unless the clinician assures confidentiality.

An interview with the family may also be fraught with difficulty when it emerges that secrets are being maintained between particular members with others, including the clinician, excluded. For instance, a close alliance between a mother and parentified child may be the locus of an intimate disclosure to the ostensible detriment of the spouse. In this situation, the clinician may well face a quandary: he surmises that the secret shared by mother and daughter are clinically pertinent; he notes the concealment's adverse effect on those kept in the dark; he has to curb his own eagerness to become privy to the secrets; and finally, he is buffeted by the realization that respect for privacy is generally a cherished ethical principle.

Remedies for these, and other, similar dilemmas are not readily forthcoming. Margolin (1982) outlines three options for the clinician working with families:

1. The confidences of members are sacrosanct and thus not to be divulged by the clinician if he is privy to them.

2. The clinician strenuously deters individual members from sharing secrets with him alone; and he is not bound by the principle of confidentiality if they are so shared (the clinician may decline to see members separately thus precluding any collusion regarding secrets (Goldenberg and Goldenberg 1991)).

3. The clinician adopts a middle position; confidentiality is not generally maintained but the patient or another family member retain the right to insist that certain information will remain private.

Apart from these options, which involve the clinician directly, the matter is further complicated during the family meeting when secrets are obviously shared by sub-groups, for example, by the parents only or by the children only. Here, the clinician could: (a) insist that his own task is made the more difficult, if not impossible, in the absence of complete disclosure; or (b) respect the boundaries set by these sub-groups groups regarding confidentiality even if this impedes the assessment process.

Most commentators on these issues share the discomforting thought that no matter how trenchantly arguments are advanced for the different options, the overall question remains unresolved. One aspect however is abundantly clear and agreed upon: whichever position the clinician adopts, he should specify explicitly its nature and potential repercussions. In seeking such informed consent, we would inform the family of our inclination for disclosure among members and to the clinician of information relevant to the purpose at hand, namely the accomplishment of the best possible diagnostic judgement and corresponding therapeutic recommendation. We would however qualify our comments by pointing out that there might be areas of an intensely personal kind whose privacy the member would wish to preserve.

Reference to informed consent brings us conveniently to our final topic: how should we go about seeking the family's agreement to participate in the clinical task.

INFORMED CONSENT

Informed consent is undoubtedly the cornerstone of ethically sound clinical practice (Redlich and Mollica 1976; Bloch and Chodoff 1991). Although the process is commonly acknowledged as warranting an important place in the clinical encounter, it is often performed perfunctorily. Moreover, consenting to an assessment procedure like a clinical interview is not viewed in the same light as a program of therapy (that is, necessitating formal permission). We contend, however, that an initial assessment, particularly one incorporating a patient and his family, is as serious and demanding an undertaking as their actual treatment and that, in any event, the boundary between assessment and treatment is necessarily blurred.

We have alluded earlier to informed consent in various ways, particularly in relation to the clinician's view of family function and dysfunction. Here, we focus specifically on its role and problems of implementation when assessing the patient and his family.

A common scenario, as described in the chapter on assessment (see Chapter 3), involves an initial meeting with a patient followed by an exploratory interview with relevant members of the person's family. Between the two points falls the first level of the informed consent process – seeking the patient's agreement to a joint meeting (we refer here to an adult or late adolescent patient, given our remit of concentrating in this volume on adult psychiatry).

The second level takes place on the telephone and/or in person when the invitees attend the clinic or hospital (it is obvious however that the telephone is a poor means of obtaining properly informed consent from a family group). Whatever the means, the ingredients of the procedure are clearly laid out; in general, they are not dissimilar from any other informed consent procedure in psychiatry where the patient is being treated voluntarily and enjoys the capacity to reflect on his needs and interests.

Following the model described by Lidz and his colleagues (1985), the following steps apply:

1. The clinician provides a comprehensive but jargon-free account of the family interview as well as of alternative clinical options, pointing out their purposes, benefits, discomforts, and risks. He also indicates the nature of his role and task and those of the family.

2. The clinician ensures that the patient in the first instance and relevant family members, subsequently, fully understand the disclosures spelt out in point 1;

3. The clinician also ensures that *all* participants consent freely, without coercion or harassment by clinician or family members. Ideally, the family is unanimous in agreement albeit with individual reservations and concerns.

4. The clinician carefully specifies that consent is required solely for the assessment interviews, *not* for a programme of therapy although a recommendation for treatment may emerge. If this is the case, a new level of consent begins.

Obvious from this outline is the notion that informed consent is a dynamic process involving a dialogue between clinician and family rather than a legalistic one-off procedure. The derivation of the term helps us to grasp its inherent nature: *con-sentire*-to feel with; the clinician and family are at one in sensing what is desirable clinically and therapeutically even though their relationship is asymmetrical, the clinician serving as the designated expert and the family accepting the need for his professionalism.

Given that a family does not necessarily think as a unit, the clinician will have to accept that one or more invited members may decline to participate. Pressure should not be exerted in this case. Indeed, the clinician must operate on the principle of respect for autonomy, acknowledging the differential interests of family members. In the event of refusal, the refuser or the remaining family members should in no way be discriminated against, for example, a veiled threat that further help will not be forthcoming, not an uncommon occurrence, should never be resorted to.

A family may of course contain young children who are not adequately equipped to grasp the information provided by the clinician. Age *per se* is not a contraindication to consent although it will be frequently the case that parents serve as proxies.

A final caveat concerns the family's vulnerability in the face of a request for a joint meeting. The scene, not difficult to imagine, is an everyday clinical one. Consider, for instance, a young adult son being treated for a relapse of his schizophrenic condition or a late adolescent in the throes of a severe episode of anorexia nervosa or a middle-aged father in a suicidal, melancholic state. The family is inevitably enveloped in a swirl of distressing feelings and confused thoughts. Given these (and many other) possibilities, the maximal amount of discretion and sensitivity is called for when seeking the family's consent for a meeting. Parents may feel blameworthy, envisaging the proposed meeting as a quasi-judicial procedure or they may anticipate a grim prognostic pronouncement. Siblings may be baffled about the patient's condition and wonder whether they too are vulnerable to a similar fate; they may also experience immense guilt about remaining unscathed compared to their severely impaired brother or sister.

An ethical guideline emerges from these reflections – *primum non nocere* – 'at first do not harm'. In his eagerness to conduct a complete evaluation, the clinician may well minimize in his own mind the potential hazards of a family gathering. As any person experienced in this area of work will testify, such a meeting is frequently charged with considerable emotional energy; this can be harnessed for good or remain unchannelled, possibly wreaking havoc. A fundamental means to achieve the former is through an acknowledgement, as part of the consent process, that a family gathering of this type is often typified by 'turbulence' (see Jenkins 1989), which is necessary for a productive exchange and useful outcome. One may also appeal to the family's altruistic impulses in stipulating that a primary purpose of the meeting is to provide an opportunity to note 'how everyone can help in the situation' (this intentional, open-endedness enables members to anticipate for themselves in what way, and for whom, help can be offered).

CONCLUSION

Our clinical experience with families and our reflections in preparing this chapter combine to reinforce a conviction that the ethical dimension of family intervention is not merely 'another' consideration when working with families but one of cardinal significance. The challenge for mental health professionals is not how to append this ethical dimension to the clinical pursuit but to appreciate their inextricable links. The gist of our viewpoint is unequivocal: the clinician's attention to ethical aspects of working with the family is not only a desideratum but also a fundamental constituent of sound clinical intervention.

Note 1

Reference to values here and by allusion in earlier sections of this chapter make it incumbent upon us to clarify what we understand by the concept and how we utilize it. This is no easy task as evidenced by the multiple approaches adopted within moral philosophy from Aristotle to the contemporary period. Both definition and theory (axiology is the term given to the theory of values) are topics of considerable debate.

A brief, schematic account will reveal the unsettled nature of the subject but also pave the way for sharing our own position (coupled with its relevance to Aponte's position concerning the optimal application of values in family work). We launch our discussion of the definition of values by stating what they are *not* – they are not concerned with what 'is', with the factual. Rather, they involve what 'ought' to be. For instance, the fact of two plus two equalling four is subject neither to preference nor desire; it is neither good nor bad; and it is not amenable to any form of evaluation. Consider by contrast an 'ought' statement: 'Parents should enhance their children's potential'. Here, the factual underpinning of the statement cannot be established; the statement reflects the author's preference; moreover, he asserts in all probability that parents ought to fulfil certain responsibilities concerning their children, regarding this as a 'good', as what is 'right'.

We cannot however infer the author's rationale or motivation from the statement as it stands. Thus, for instance, the author may be a politician proclaiming that enhancement of childrens' potential can never be a state, bureaucratic function but must remain a responsibility of parents. Alternatively, it could be a preference espoused by a family therapist contending that family workers should, as a primary goal, support parents in order to maximize their wherewithal to do the best by their children. Whatever the author's motivation, we can conclude at the least that the statement is his preference for what ought to be.

It still is the case that what ought to be is liable to interpretation. For R. M. Hare (1952), the Oxford moral philosopher, value refers to what is right or wrong, good or bad, desirable or undesirable. Therefore, making a value judgement reflects how a person appraises an issue, event, statement, or whatever, in terms of these dimensions. For other philosophers, value is viewed more broadly, covering not only these dimensions but also aspects like virtue, truth, beauty, and obligation (Frankena 1967).

Although substantially varied, theories of value illuminate the subject further. In Frankena's useful summary, these theories are grouped under the categories of normative and meta-normative. Normative theories tackle the question of what should be regarded as an intrinsic value. For example, should the criterion be what is pleasant or what amounts to personal flourishing (for example, Aristotle's *eudaemonia*), or what is satisfying or what is self-realizing, or what is aesthetic or what is virtuous, and so on – the list can be extended indefinitely.

Meta-normative theories are not preoccupied with actual criteria but focus instead on the *nature* of value. For example, a value judgement may be regarded as descriptive inasmuch as it invests definable properties onto objects or events or experience (say a property of being enjoyed or being desired). Alternatively, a value is held to be a metaphysical property, not liable to empirical examination (say something willed by God). Another view, the intuitionist, states the indefinable quality of a value – beyond natural or empirical domains; the claim is that a value inheres in objects independent of whether they are desired or enjoyed. Yet another position suggests that value judgements should be regarded as prescriptions, that is as recommending a specific, ethically determined course of action. A final view, contrary to the prescriptivist, highlights the inevitable arbitrariness of all value judgements and claims that they are impervious to any justificatory process (Sartre and the European Existentialists are representive).

REFERENCES

Aponte, H.J. (1985). The negotiation of values in therapy. *Family Process*, **24**, 323–38.

Asen, K., Berkowitz, R., Cooklin, A. *et al.* (1991). Family therapy outcome research: a trial for families, therapists, and researchers. *Family Process*, **30**, 3–20.

Bentovim, A. (1989). Systems mysticism and child sexual abuse – a rebuttal. *Context*, **2**, 22–4.

Bentovim, A., Elton, A., Hildebrand, J., Tranter, M., and Vizard, E. (1988). *Child sexual abuse within the family*. Wright, London.

Bloch, S. (1989). The student with a writing block – the ethics of psychotherapy. *Journal of Medical Ethics*, **15**, 153–8.

Bloch, S. and Chodoff, P. (ed.) (1991). *Psychiatric ethics*. Oxford University Press, Oxford.

Bloch, S., Sharpe, M., and Allman, P. (1991). Systemic family therapy in adult psychiatry: a review of 50 families. *British Journal of Psychiatry*, **159**, 357–64.

Boszormenyi-Nagy, I., Grunebaum, J., and Ulrich. D. (1991). Contextual therapy. In *Handbook of family therapy, vol. 2* (ed. A.S. Gurman and D.P. Kniskern). Brunner/Mazel, New York.

Carr, A. (1989). Chloe Madanes' comprehensive guide to strategic therapy. *Context*, **2**, 11–14.

Doherty, W. and Boss P. (1991). Values and ethics in family therapy. In *Handbook of family therapy, Vol. 2* (ed. A.S. Gurman and D. Kniskern). Brunner/Mazel, New York.

Fieldsteel, N.D. (1982). Ethical issues in family therapy. In *Ethics and values in psychotherapy: a guide-book* (ed. M. Rosenbaum). Free Press, New York.

Frankena, W.K. (1967). Value and valuation. In *The encyclopedia of philosophy* (ed. P. Edwards). Macmillan, New York.

Goldenberg, I. and Goldenberg, H. (1991). *Family therapy: an overview* 3rd ed. Brooks/Coles, Pacific Grove, CA.

Goldner, V. (1985). Feminism and family therapy. *Family Process*, **24**, 31–47.

Goldner, V. (1988). Generation and gender: Normative and covert hierarchies. *Family Process*, **27**, 17–31.

Hare, R.M. (1952). *The language of morals*. Oxford University Press, Oxford.

Holmes, J. and Lindley, R. (1989). *The values of psychotherapy*. Oxford University Press, Oxford.

Jenkins, H. (1989). Precipitating crisis in families: patterns which connect. *Journal of Family Therapy*, **11**, 99–109.

Kant, E. (1973). *Critique of pure reason*. Macmillan, London.

Lakin, M. (1988). *Ethical Issues in the psychotherapies*. Oxford University Press, New York.

Leupnitz, D. (1988). *The family perspective: feminist theory in clinical practice*. Basic Books, New York.

Lidz, C.W., Meisel, A., Zerubavel, E., Carter, M., Sestak, R.M., and Roth, L.H. (1984). *Informed consent: a study of decision-making in psychiatry*, Guilford, New York.

Lustig, N., Dresser, J. W., Spellman, S., and Murray, T. (1966). Incest: a family group survival pattern. *Archives of General Psychiatry*, **14**, 31–40.

Margolin, G. (1982). Ethical and legal considerations in marital and family therapy. *American Psychologist*, **37**, 788–801.

Palazzoli, M., Boscolo, L., Cecchin, G., and Prata, G. (1980). Hypothesizing, circularity, neutrality: Three guidelines for the conductor of the session. *Family Process*, **19**, 3–12.

Redlich, F. and Mollica, R.F. (1976). Overview: Ethical issues in contemporary psychiatry. *American Journal of Psychiatry*, **133**, 125–36.

Steere, J. and Dowdall, T. (1990). On being ethical in unethical places: The dilemmas of South African clinical psychologists. *Hastings Center Report*, **20**, 11–15.

Waldegrave, C., Campbell, W., and Tamasese, K. (1991–92). Just therapy. *Context*, **10**, 32–3.

Walrond-Skinner, S. and Watson, D. (1987). *Ethical issues in family therapy*. Routledge and Kegan Paul, London.

Wendorf, R. and Wendorf, D. (1985). Rejoinder: looking high and low. *Family Process*, **24**, 457–60.

Will, D. (1989). Feminism, child sexual abuse and the (long overdue) demise of systems mysticism. *Context*, **1**, 12–15.

Index